Virtue Ethics and the Environment

This book addresses one of today's most burning issues, namely the environmental crisis, by offering an insight into the problem from the perspective of virtue ethics.

Virtue ethics is an approach to ethics that centralizes the concept of moral virtue, which can be extended to environmental ethics via environmental virtue ethics (EVE). Beginning with a comprehensive overview, the book explores the renaissance of contemporary virtue ethics and the beginnings of EVE in the second half of the 20th century and presents the main characteristics, proponents, and criticisms of EVE. The book then goes on to analyze its development by distinguishing the three most influential concepts: the classical; the naturalistic, teleological, and pluralistic; and the narrative conception of environmental virtue ethics. The author also discusses the most influential works on EVE, including a revision of Louke van Wensveen's postulate to use virtue language in environmental ethics. By synthesizing such works on EVE alongside an analysis of the three most important concepts, the book offers a new concept that is universalistic, positive, and pragmatic.

The book will be useful for students, scholars, and researchers studying environmental ethics, sustainable development, environmental psychology, moral philosophy, and philosophy of education.

Dominika Dzwonkowska is Professor of Philosophy in the Institute of Philosophy at Cardinal Stefan Wyszynski University in Warsaw, Poland.

Routledge Environmental Ethics
Series Editor: Benjamin Hale
University of Colorado, Boulder

The Routledge Environmental Ethics series aims to gather novel work on questions that fall at the intersection of the normative and the practical, with an eye toward conceptual issues that bear on environmental policy and environmental science. Recognizing the growing need for input from academic philosophers and political theorists in the broader environmental discourse, but also acknowledging that moral responsibilities for environmental alteration cannot be understood without rooting themselves in the practical and descriptive details, this series aims to unify contributions from within the environmental literature.

Books in this series can cover topics in a range of environmental contexts, including individual responsibility for climate change, conceptual matters affecting climate policy, the moral underpinnings of endangered species protection, complications facing wildlife management, the nature of extinction, the ethics of reintroduction and assisted migration, reparative responsibilities to restore, among many others.

For more information on the series, please visit https://www.routledge.com/ Routledge-Environmental-Ethics/book-series/ENVE

Virtue Ethics and the Environment

Dominika Dzwonkowska

Ministry of Science and Higher Education
Republic of Poland

Routledge
Taylor & Francis Group
LONDON AND NEW YORK

earthscan
from Routledge

First published 2025
by Routledge
4 Park Square, Milton Park, Abingdon, Oxon OX14 4RN

and by Routledge
605 Third Avenue, New York, NY 10158

Routledge is an imprint of the Taylor & Francis Group, an informa business

The publication is supported by state budget funds granted by the Ministry of Science and Higher Education in Poland within the framework of the Excellent Science Program. Project title: Ethics of Environmental Virtues, MONOG/SP/0223/2023/01.

Translated from the Polish language edition by Dominika Dzwonkowska:
Etyka cnót środowiskowych (Eng: Environmental Virtue Ethics) by Dominika Dzwonkowska
Published 2019 by Cardinal Stefan Wyszyński University's Publishing House
ISBN: 978-83-8090-638-9

British Library Cataloguing-in-Publication Data
A catalogue record for this book is available from the British Library

ISBN: 978-1-032-55970-4 (hbk)
ISBN: 978-1-032-55971-1 (pbk)
ISBN: 978-1-003-43315-6 (ebk)

DOI: 10.4324/9781003433156

Typeset in Times New Roman
by SPi Technologies India Pvt Ltd (Straive)

Contents

Introduction

The word "virtue" refers to a certain aspect of moral life and has played an important role in philosophy almost since the very beginning of ethical reflection. The term has undergone many semantic transformations corresponding to trends in thinking about morality. Its meaning has been constantly adapted to the needs of the time and to cultural and historical contexts, taking into account how ethics responds to the moral challenges facing man. Today, the environmental crisis appears to be a tremendous challenge for human beings. From the ethical perspective, the moral dimension of human–environmental relations is crucial. The moral aspect of the environmental crisis has been recognized since the 1960s and has been analyzed from the perspectives of various ethical traditions and of virtue, which led to the beginning of environmental virtue ethics (EVE).

Environmental virtue ethics is a relatively young area of ethical consideration, but it makes an important contribution to the discussion of the moral agent's duties toward the environment. First of all, it reflects on ecology from a completely different perspective than other ethics, directing its main attention to the hitherto somewhat neglected moral agent. It shows the nature of ethical obligations from the perspective of the moral agent and his aspiration to become the best possible version of himself, primarily including his state of ethical perfection or happiness (eudaimonia). In this way, environmental virtue ethics links a sense of fulfillment in life with a concern for the realization of moral obligations to the natural environment. Moreover, it addresses one of the most pressing problems, namely the environmental crisis. Thus, environmental virtue ethics brings out the depth of ancient wisdom and shows that, despite the passage of time, it seems to be the right approach for solving problems faced by the moral agent today.

In this book, I address the application of virtue ethics to the debate over the state of the environment, seeking answers to specific questions and problems that arise in EVE discussion: what is the place of the environment and environmental protection in the realization of moral dispositions? Is protection of the environment a *sine qua non* for the full ethical development of man? What is environmental virtue and vice? What effect do environmental virtues and vices have on the moral agent in the context of his excellence? How (if at all) do

DOI: 10.4324/9781003433156-1

environmental virtues and vices differ from other virtues and vices? Can environmental virtue ethics contribute to solving the problem of environmental degradation?

The purpose of this monograph is to outline the concept of a universalist, positive, and practical ethics of environmental virtues. This goal will be achieved in two stages. First, by showing the panorama of aretological thought in contemporary ethics and environmental virtue ethics. The first two parts of the book are devoted to presenting the existing concepts. The first part introduces the issue of virtue ethics and provides a canvas for further consideration and critical analysis of environmental virtue ethics. In this part of the monograph, I will present contemporary virtue ethics and environmental virtue ethics. I will also show the concept of the so-called language of virtues introduced by Louke van Wensveen, and in Section 8.2 I propose a correction to this Dutch researcher's postulate.

In the second part of the book, I will present three autonomous concepts of environmental virtue ethics: the classical conception of environmental virtue ethics (Henry David Thoreau); the naturalistic, teleological, and pluralistic conception of environmental virtue ethics (Ronald Sandler); and the narrative conception of environmental virtue ethics (Brian Treanor).

The second stage of my research goal is to show the outline of a universalist, positive, and practical conception of environmental virtue ethics. In the third part of the book, I claim that environmental virtue ethics should take into account the universal, positive, and practical nature of virtues. Moreover, the practical dimension of virtues requires adapting philosophical concepts to the practical nature of virtue. It is necessary to expand the language of virtues to include philosophical concepts that take into account the realm of praxis. Thus, I will use Mark Coeckelbergh's concept of environmental skills to propose a correction of van Wensveen's postulate regarding virtue language.

I base the monograph mainly on Anglo-Saxon literature, since environmental virtue ethics is a rather narrow discipline, being primarily in the area of interest of environmental ethicists publishing in English. Therefore, in order to better understand this discourse, I draw on works published in the US and the cultural context of the environmental tradition in American culture. From this perspective, the works of several authors are crucial, primarily the most important environmental virtue ethicists, including Henry David Thoreau, Philip Cafaro, Ronald Sandler, Louke van Wensveen, Brian Treanor, Geoffrey Frasz, Thomas Hill, Louke van Wensveen, Holmes Rolston III, John O'Neil, Thomas Hill Jr., Lisa Newton, Rosalind Hursthouse, Val Plumwood, Allen Thompson, and Jason Kawall. The basis for understanding the American discussion of environmental ethics should also be sought in the texts of the most important authors in this area, such as Aldo Leopold and Paul Taylor.

However, I owe the perspective on ecological discourse in philosophy to Polish philosophers who are representatives of the so-called first wave of environmental ethics (cf. Dzwonkowska 2017). These are primarily Włodzimierz Tyburski, Zdzisława Piątek, Józef M. Dołęga, Andrzej Papuzinski, Marek

Bonenberg, and, somewhat later, Helena Ciążela. The revival of virtue ethics that took place in the second half of the 20th century is an interesting area of research that is also undertaken by Polish thinkers. It is worth giving at least a few names of researchers who have contributed to the analysis of contemporary virtue ethics: Jacek Jaśtal, Natasha Szutta, Artur Szutta, Marcin Zdrenka, and Piotr Domeracki. Their works not only present original research concepts but also play a significant role in spreading awareness in Poland of current English-language research.

The virtue debate itself is a valuable element of ethical discourse, first and foremost supplementing the ongoing exchange of ideas by demonstrating the importance of the moral agent in taking action toward the ethical good. As was the case with the two dominant currents of Enlightenment ethics, the virtue debate detaches ethical discussion from debating the act or its consequences, thus directing attention to the moral agent itself and its role in ethics. Analogically to this, environmental virtue ethics complements considerations of humans' moral obligations to the environment by asking what kind of human wants to protect the environment and what kind destroys it.

References

Dzwonkowska D., *Environmental Ethics in Poland*, "Journal of Agricultural and Environmental Ethics" 2017, vol. 30, p. 135–151.

Part I

Introduction to environmental virtue ethics

1   Contemporary virtue ethics

Virtue ethics is one of the oldest concepts, dating back to ancient thought, so it would be difficult to include the richness of the virtue tradition and the multitude of approaches to aretological theory in a single monograph. Hence, this chapter mainly presents virtue ethics in contemporary writing, following the so-called renaissance of virtue ethics that took place in the second half of the 20th century. The renaissance of virtue ethics is attributed to Gertrude Elizabeth Margaret Anscombe (1958), who in her article *Modern Moral Philosophy* advocates an end to discussions of whether an act is good or bad, instead proposing a shift of focus to the moral agent. Anscombe accuses dominant approaches to ethics of being ineffective in solving ethical problems and unable to contribute to humanity's moral progress. The moral theories she criticizes focus on trying to show whether a deed is right on the basis of the consequences arising from it (consequentialism), or they judge a deed on the basis of its conformity to norms, codes, or prohibitions (deontology). The various proponents of consequentialism agree that "the right action can be defined in terms of good consequences" (Van Zyl 2019, 3). While for deontologists this argument regarding consequences is not acceptable, they tend to consider a deed as morally good only when it is in accordance with moral rules. Thus, a question arises about the source of moral duty, which could be God's commands, reason (a tradition started by Kant) or intuition (from W.D. Ross) (cf. Van Zyl 2019).

Virtue ethics has a different perspective on the nature of moral right and wrong. Van Zyl points out that Anscombe criticizes the incoherence of the concepts of moral obligation, duty, and right and wrong – terms that are derived from Judeo-Christian theory and presuppose the existence of a law-giver of universal moral law. However, over time "the idea that such a law-maker existed was rejected, and yet people continued to use the concepts of moral duty and right and wrong action" (2019, 6). Anscombe in her article (1958) claimed that the discussion of moral duty and morally right or wrong should be replaced by a discourse on virtue and vice. This changed the paradigm of moral thinking: instead of a specific moral code, catalogs of norms, or analysis of the consequences of an action, ethics returns to the ancient tradition that focuses on a moral agent's dispositions. A person is supposed

DOI: 10.4324/9781003433156-3

to act in accordance with their virtue, to do good, guided by an inner disposition and practical wisdom rather than by a certain moral norm.[1] Van Zyl
claims that the advantage of

> virtue and vice terms is that they are not only evaluative but also descrip-
> tive. Advising someone to do what is generous, honest, or courageous is
> much more informative than merely advising them to do what is right: it
> directs their attention to the situation.
>
> (2019, 7)

Van Zyl also points out that virtues and vices rely on character traits that exist
in the real world – not on metaphysical entities like a divine lawgiver or a set of
moral duties (Van Zyl 2019).

The first articles in the field of contemporary virtue ethics marked its peculiarity and difference from the dominant ethical traditions in the ethical debate,
namely deontological and consequentialist ethics. It should be emphasized
that both deontological and consequentialist ethics include virtues, consequences, and principles in their approaches; however, they define them differently and assign them a different role in ethical theory.

> Consequentialists will define virtues as traits that yield good conse-
> quences and deontologists will define them as traits possessed by those
> who reliably fulfil their duties, virtue ethicists will resist the attempt to
> define virtues in terms of some other concept that is taken to be more
> fundamental. Rather, virtues and vices will be foundational for virtue
> ethical theories and other normative notions will be grounded in them.
>
> (Hursthouse, 2016)

The main focus of virtue ethics is the traits of a moral agent, namely virtues
and vices that

> are relatively stable dispositions to act in a certain way (…). Virtues are
> good character traits, and vices are bad character traits. We praise and
> admire people who are honest, kind, just generous, courageous. (…) we
> blame and criticize people for being dishonest, unkind, selfish, or
> arrogant
>
> (van Zyl 2019, 9)

The renaissance of virtue ethics has revived the discussion in contemporary philosophy on a moral agent's dispositions by introducing the concept of aretological issues. From a theoretical point of view, it is therefore necessary to distinguish
virtue ethics from virtue theory.[2] The latter term denotes a particular conception
of virtues in deontological or consequentialist ethics (Hursthouse 2016) or
refers to any theoretical discussion of the nature of virtues and vices, even if the
role of virtue theory is not central in a given ethical system (Jost 2005, 679).

Contemporary virtue ethics is very diverse and is visible in a variety of ways of dealing with aretological issues, but some see it mostly as a continuation of ancient thought; according to Jacek Jaśtal, "in terms of positive solutions, however, this ethics has not really gone beyond (…) extensive reinterpretations of ancient writers" (2004, 38). References to eudaimonism in terms inspired by Aristotelian thought predominate in contemporary approach (e.g., Philippa Foot or Rosalind Hursthouse), although Michael Slote also has conceptions that distance themselves from eudaimonism.

Even though the theoretical framework of virtue ethics was outlined by Aristotle, it should not be thought that it is associated only with the philosophy of this Stagirite, since a discussion on virtues runs through all ancient ethical theories (Annas 1993). The modern reception of virtue ethics remains most often within the Aristotelian framework, but Aristotle's thought is sometimes freely interpreted. In addition, many philosophers seek alternative interpretations of virtue ethics in other philosophical conceptions, such as in the thought of Francis Hutcheson, David Hume, Friedrich Nietzsche, James Martineau, or Martin Heidegger (Hursthouse 2016). Many philosophers also refer to inspirations from outside the Western cultural circle and reach back to Far Eastern traditions, interpreting virtue ethics through the prism of Confucianism or Buddhist or Hindu philosophies.[3] In addition, the aretological approach has become an area eagerly exploited by applied ethics, hence there are attempts to apply virtue ethics to business ethics,[4] medical issues,[5] technological innovations,[6] or environmental problems. Such diverse approaches open new areas for discussion of virtue ethics and ways of perceiving it. Above all, however, they are an attempt to seek answers to the challenges of modern times in clear ethical theories.

1.1 The main trends of modern virtue ethics

According to Rosalind Hursthouse (2016), contemporary concepts of virtue ethics can be divided into four distinct types, in each of which the concept of virtue is a central issue, but the understanding of virtues and their place varies. Thus, one can distinguish between (a) eudaimonist virtue ethics; (b) agent-based and exemplarist virtue ethics; (c) target-centered virtue ethics; and (d) Platonistic virtue ethics. Each of these concepts addresses the issue of exemplary human moral behavior and takes into account the practical wisdom necessary to take ethically right actions. Nonetheless, they are differently formulated and differ in their selection of key elements of virtue ethics, its goals, and its modes of application.

Eudaimonist virtue ethics relates primarily to eudaimonia as the goal of ethical life, linking moral progress to the attainment of lasting happiness and a high quality of life for the moral agent. Many thinkers of the ancient world shared the view that good, virtuous conduct would bring lasting happiness to man and guarantee him a happy life. Eudaimonism could be defined as an ethical view that proclaims that the pursuit of happiness is the driving force behind human actions and their highest goal. Thus, "the final end for the sake of which we pursue all other goals, then, is true happiness or *eudaimonia*" (van Zyl 2015, 185). Happiness can be guaranteed by appropriate conduct, such as the path of virtue, as was believed by Socrates and Aristotle; the path of *apatheia*, according to the Stoics; or pleasure, according to the Epicureans. Parry points out several possible ways of understanding eudaimonism (Parry 2014): a) virtue, together with its active manifestation, is identical to happiness; b) virtue, together with its active manifestation, is the most important component of happiness; c) virtue is only a means to achieve happiness. At the same time, Parry cautions against the simplistic statement that ancient theories base the value and importance of virtue only on the fact that it is a means to achieve happiness.

Eudaimonia is thus a moralized conception of a happy life (Swanton 2003, 87–90). Above all, happiness itself is not a superficial feeling of excitement or contentment but a certain form of fulfillment and the attainment of lasting satisfaction (cf. Prior 2001). The sources of this happiness in virtuous people are different from those of non-virtuous people, for happiness comes from actions guided by appropriate virtues. Christine Swanton (Swanton 2003, 87–90) gives an example of the virtue of friendship, so highly valued by Aristotle. As Swanton argues, if friendship serves only to complain and whine, it is harmful to both friends. So even in friendship, it is important that it is guided by the right motivation and motives. The same is true of happiness: a virtuous person will obtain it in a manner consistent with his virtuous endowment.

Many ethical concepts contrast eudaimonia with the happiness that comes from material possessions or physical pleasure. Eudaimonia is a form of happiness that transcends the satisfaction of any kind of sensual pleasure. According to Aristotle, virtue is necessary for a happy life, but external goods are also important. Plato and the Stoics, on the other hand, held that the possession of virtue is not only necessary but also sufficient for a good life, and that material goods are not a condition for eudaimonia (Cf. van Zyl 2015, 183; 191–192; McDowell 1980, 359–376).

The second type of virtue ethics focuses on the moral agent and exemplary moral character. In this view, a moral agent and his motivations and character traits (dispositions) are of particular importance. This view of virtue ethics refers to the ethical ideal in order to identify the qualities of the moral agent that are important in the pursuit of a truly happy life. The sources of normativity are thus sought in the motivations and qualities of the moral agent. Michael Slote (2001: 14) believes that the moral qualification of an act depends on the motivation of the moral agent. A good act is the fruit of the good motives that guide an individual, whereas a bad act is the fruit of bad (or insufficiently good) motives (Hursthouse 2016). According to Slote, the goodness of an act, the fairness of a law or a social institution, and the normativity of practical rationality depend on the motivation and disposition of the moral agent. In Linda Zagzebski's view, the goodness of an act should be defined in relation to the emotions, motives, and dispositions of a moral agent who is virtuous or has moral defects. For example, a bad act is one that a *phronimos* would ordinarily not perform or would feel guilty about if it were committed (Hursthouse 2016). The moral qualification of an act is determined by the motivations, dispositions, and emotions of the moral agent. As Zagzebski claims,

> A wrong act = an act that the *phronimos* characteristically would not do, and he would feel guilty if he did = an act such that it is not the case that he might do it = an act that expresses a vice = an act that is against a requirement of virtue (the virtuous self).
> (Zagzebski 2004, 160; cf. Hursthouse 2016)

According to Slote, the motivation and disposition of the moral agent during the performance of the act being evaluated are crucial in determining how motivation and disposition relate to explaining other normative characteristics. "The goodness of action A, for example, is derived from the agent's motives when she performs A" (Hursthouse 2016). If the moral agent is guided in his action by good motives, the act is good; if there is no good motivation, the act cannot be considered good. Linda Zagzebski does not pay attention to the moral agent's motives but rather asks whether a given act is the kind of act that a virtue-driven moral agent would perform. "Appealing to *the virtuous agent's hypothetical motives and dispositions* enables Zagzebski to distinguish between performing the right action and doing so for the right reasons (a distinction that, as Brady (2004) observes, Slote has trouble drawing)" (2004).

Another important issue in the subject under discussion is the question of identifying moral role models or moral exemplars: how and on the basis of which principles we distinguish between those whom we admire for their behavior and those who do not meet the criteria of a moral role model or even commit behavior considered morally reprehensible. Zagzebski (2004, 41) emphasizes that we do not know the criteria of goodness before we identify individuals whose behavior we consider exemplary. According to her, our understanding of the motives and dispositions of the moral agent is based on simple observation of everyday life. By peeping at people in our own surroundings, we perceive that some of them behave in ways we want to imitate, and it is they who provide us with positive moral role models, while the behavior of others represents negative moral role models and therefore does not arouse in us the desire to imitate. This contact with role models and our reaction to them build up our understanding of the motives and disposition of a virtuous person or one who is morally flawed. Our comprehension is made more complete by observing various role models.

The third approach is target-centered virtue ethics, developed by Christine Swanton (ibid.; cf. Swanton 2003). This view emphasizes that a full description of a virtue will include "1) its *field*, 2) its *mode* of responsiveness, 3) its *basis* of moral acknowledgment, and 4) its *target*" (Hursthouse 2016).

Various virtues are within different *fields*. "Courage, for example, is concerned with what might harm us, whereas generosity is concerned with the sharing of time, talent, and property" (Hursthouse 2016). Virtue's *mode* is about how a virtue addresses the basis of moral acknowledgment within its field; for example, "generosity promotes a good, namely another's benefit, whereas courage defends a value, bond, or status" (Hursthouse 2016). The basis of acknowledgment

> of a virtue is the feature within the virtue's field to which it responds (...) generosity is attentive to the benefits that others might enjoy through one's agency, and courage responds to threats to value, status, or the bonds that exist between oneself and particular others, and the fear such threats might generate
>
> (Hursthouse 2016)

Virtue's target is the end toward which virtue leads a moral agent. Thus, a "*virtuous act* is an act that hits the target of a virtue, which is to say that it succeeds in responding to items in its field in the specified way" (Hursthouse 2016, Cf. Swanton 2003, 233).

The fourth account of virtue ethics comes from the Platonic tradition. In a number of Platonic dialogues, Socrates[7] attempted to define the essence of individual virtues. These took the form of discussions with philosophers or citizens of Athens of his times. In the course of these discussions, attempts were made to demonstrate the fallacy of the interlocutor's beliefs and then to guide him to the correct understanding of the virtue under discussion. Many virtue ethicists take the Platonic dialogues as a starting point and, using them, proceed to defend a *eudaimonistic* virtue ethic. Hursthouse distinguishes two crucial trends in virtue ethics in the Platonic spirit (Hursthouse 2016).

The first of these trends is characteristic of Timothy Chappell (2014), who follows the views of Iris Murdoch (1971). Chappell, in an attempt to answer the question of what virtue is in terms of Platonic dialogues, states that it always presupposes the prior contemplation of the Form of Good and calls "fat relentless ego" the enemy of virtue (1971). Contemplation of the Good is a prerequisite for being able to act well, because all our needs, desires, thoughts, and passions distort our perception, preventing us from seeing the good that is around us. Regular contemplation of the good sensitizes us to what is important and helps us to be less subject to the factors that distort our perceptions. Contemplation leads to getting rid of subjectivism and egoism; therefore it is a virtue. Thus, virtue "pierces the veil of selfish consciousness and joins the world as it really is" (1971).

Robert Merrihew Adams (1999) offered another Platonic account of virtue ethics whose starting point is metaphysical findings about the nature of the good. Following St. Augustine, he regards God as the highest Good. According to him, God is the source and example of all goodness; every other thing can be considered good only to the extent that it resembles God. Goodness is fully realized only in the form of the personal God. The question then arises whether a person or action can be considered virtuous. "Virtues come into the account as one of the ways in which some things (namely, persons) could resemble God" (1999). Thus, virtuous deeds are one of the ways in which a person can become like God. Adams recognized that a personal God is a better example of perfection than impersonal goodness.

Virtue ethics has sparked keen interest in aretological issues. The few names already mentioned do not reflect the richness of the discussion conducted in this area. Worth mentioning, for example, is Alasdair MacIntyre, who, because of his different style of philosophizing, is sometimes overlooked in discussions conducted by the most important contemporary virtue ethicists. Nevertheless, his work on virtue ethics is an important and very well-argued voice in the aretological discussion. The starting point for this British philosopher is the recognition of the helplessness of modern ethics in overcoming moral disputes; this is due to the incommensurability of various ethical positions, the inadequacy of concepts, and the deceptiveness of the modern concept of morality and moral relativism (Machura 2002; Gałkowski 2004). Emotivism becomes the main defendant, and the antidote to this condition can only be a return to virtue ethics. The characteristic feature of this conception of virtue is marked by MacIntyre's communitarian views (Kuniński, 2006), according to which the key element in the formation of virtue is the social dimension. On the one hand, the way in which virtues are perceived and understood is influenced by the moral agent; on the other hand, virtues are influenced by the space in which the moral agent develops them. MacIntyre repeatedly emphasizes that virtues arise in a particular community and are an expression of some tradition. As an element of this tradition, they are subject to change along with the tradition itself, which entails transformations in the way morality is understood. MacIntyre's position is very well argued and extremely well presented.

1.1.1 Summary

Virtue ethics has seen a resurgence in modern philosophy as an expression of dissatisfaction with the ethical solutions proposed by the dominant modern ethical approaches, namely deontological ethics and utilitarian ethics. It has been recognized that the ineffectiveness of ethics in answering important moral questions could be resolved by redirecting attention from the act and its consequences to the moral agent. With the development of modern virtue ethics, the initial criticism of other ethical currents gave way to attempts to see the theoretical problems they shared. Over time, aretological theories were restored, even revisiting many ethical views that were initially criticized. Virtue ethics itself has become an important area of moral philosophy.

Modern virtue ethics makes abundant use of Aristotle's ethics, mainly the concept of eudaimonism. Often, however, interpretations of this Stagirite's thought are quite free. Other inspirations also appear, such as references to

Plato, the Stoics, David Hume, Friedrich Nietzsche, or Martin Heidegger. One can speak of four distinct ways of practicing virtue ethics: (a) eudaimonist virtue ethics; (b) agent-based and exemplarist virtue ethics; (c) target-centered virtue ethics; and (d) Platonistic virtue ethics. Each of these approaches addresses the quandary of exemplary human moral qualities.

Notes

1 It should be noted that some treat virtue ethics as the opposite of deontological and utilitarian traditions, while others treat the former as complementary to the latter (cf. Sandler 2018, p. 226–232).

2 In environmental virtue ethics, often, in order to avoid answering certain theoretical problems, the term "virtue-oriented ethics" is used in place of "environmental virtue ethics" (cf. Sandler 2007).

3 This topic has recently gained more attention, and there are a few important contributions that should be mentioned here: D. E. Cooper, S. P. James, *Buddhism, virtue, and environment*, New York 2017; Perret R. W., G. Pettigrove, *Hindu Virtue Ethics*, in: *The Routledge Companion to Virtue Ethics*, L. Besser-Jones, M. Slote (ed.), New York 2015, p. 51–62; M. Sim, *Why Confucius' ethics is a virtue ethics*, in: *The Routledge Companion to Virtue Ethics*, L. Besser-Jones, M. Slote (ed.), New York 2015, p. 63–76; Ch. Goodman, *Virtue in Buddhist ethical tradition*, in: *The Routledge Companion to Virtue Ethics*, L. Besser-Jones, M. Slote (ed.), New York 2015, p. 89–98; Huang Y, *Respect for Differences: The Daoist Virtue*, in: *The Routledge Companion to Virtue Ethics*, L. Besser-Jones, M. Slote (ed.), New York 2015, p. 99–112; E. L. Hutton, *Xunzi and virtue ethics*, in: *The Routledge Companion to Virtue Ethics*, L. Besser-Jones, M. Slote (ed.), New York 2015, p. 113–125; Y. Xiao Y., *Virtue Ethics as Political Philosophy: The Structure of Ethical Theory in Early Chinese Philosophy*, in: *The Routledge Companion to Virtue Ethics*, L. Besser-Jones, M. Slote (ed.), New York 2015, p. 471–489.

4 For example: R. Audi, *Business Ethics from a Virtue-Theoretic Perspective*, in: *The Routledge Companion to Virtue Ethics*, L. Besser-Jones, M. Slote (ed.), New York 2015, p. 529–542; J.L. Nicholas, G. Moore, *Virtue at Work: Ethics for Individuals, Managers, and Organizations*, "Philosophy of Management" 2018, no. 17, p. 257–259; E. Hartman, *Virtue Ethics and Business: An Aristotelian Approach*, Cambridge 2013; Richards D. G., *Economics, Ethics, and Ancient Thought*, New York 2017.

5 For example, R. L. Walker, *Virtue Ethics and Medicine*, in: *The Routledge Companion to Virtue Ethics*, L. Besser-Jones, M. Slote (ed.), New York 2015, p. 515–528; Baresford E.B., *Can phronesis Save the Life of Medical Ethics?*, "Theoretical Medicine and Bioethics" 1996, vol. 17, p. 209–224; T. S. Huddle, *Teaching Professionalism: Is Medical Morality a Competency?*, "Academic Medicine" 2005, vol. 80, no. 1, 885–891; Jansen L.A., *The Virtue in the Place: Virtue Ethics in Medicine*, "Journal of Theoretical Medicine and Bioethics" 2000, vol. 21, p. 261–276; E.D. Pellegrino, *Towards a Virtue Based Normative Ethics for the Health Professions*, "Kennedy Institute of Ethics Journal 1995, vol. 5, no. 3, p. 253–277.

6 For example, Coeckelbergh M., *Environmental Skill. Motivation, Knowledge, and the Possibility of a Non-Romantic Environmental Ethics*, New York 2015; Gunkel D. J., *The Machine Question. Critical Perspectives on AI, Robots, and Ethics*, Cambridge MA 2012; Vallor S., *Technology and the Virtues. A Philosophical Guide to a Future Worth Wanting*, Oxford 2016.

7 The dialogues Hursthouse writes about are Plato's early dialogues, the so-called Socratic dialogues. They are an important source of knowledge about Socrates, and

especially about the maieutic method he used, whose purpose was to extract knowledge from the interlocutor by asking questions.

References

Adams R. M., *Finite and Infinite Goods*, New York 1999.

Annas J., *The Morality of Happiness*, Oxford 1993.

Anscombe G. E. M., *Modern Moral Philosophy*, "Philosophy" 1958, vol. 33, no. 124, p. 1–19.

Audi R., *Business Ethics from a Virtue-Theoretic Perspective*, in: *The Routledge Companion to Virtue Ethics*, L. Besser-Jones, M. Slote (ed.), New York 2015, p. 529–542.

Baresford E. B., *Can Phronesis Save the Life of Medical Ethics?*, "Theoretical medicine and bioethics" 1996, vol. 17, no. 3, p. 209–224.

Brady M. S., *Against Agent-Based Virtue Ethics*, "*Philosophical Papers*" 2004, vol. 33, no. 1, p. 1–10.

Chappell T., *Knowing What to Do*, Oxford 2014.

Coeckelbergh M., *Environmental Skill. Motivation, Knowledge, and the Possibility of a Non-Romantic Environmental Ethics*, New York 2015.

Cooper D. E., S. P. James, *Buddhism, Virtue, and Environment*, New York 2017.

Gałkowski S., *Cnoty i relatywizm. Alasdaira MacIntyre'a próba przekroczenia relatywizmu*, "Diametros" 2004, vol. 1, no. 2, p. 1–17.

Goodman C., *Virtue in Buddhist Ethical Tradition*, in: *The Routledge Companion to Virtue Ethics*, L. Besser-Jones, M. Slote (ed.), New York 2015, p. 89–98.

Gunkel D. J., *The Machine Question. Critical Perspectives on AI, Robots, and Ethics*, Cambridge MA 2012.

Hartman E., *Virtue Ethics and Business: An Aristotelian Approach*, Cambridge MA 2013.

Huang Y., *Respect for Differences: The Daoist Virtue*, in: *The Routledge Companion to Virtue Ethics*, L. Besser-Jones, M. Slote (ed.), New York 2015, p. 99–112.

Huddle T. S., *Teaching Professionalism: Is Medical Morality a Competency?*, "Academic Medicine" 2005, vol. 80, no. 10, p. 885–891.

Hursthouse R., *Virtue Ethics*, w: *Stanford Encyclopedia of Philosophy*, E.N. Zalta (ed.) (Winter 2016 edition), 2016, available at: https://plato.stanford.edu/archives/win2016/entries/ethics-virtue (Access: December 15, 2022).

Hutton E. L., *Xunzi and Virtue Ethics*, in: *The Routledge Companion to Virtue Ethics*, L. Besser-Jones, M. Slote (ed.), New York 2015, p. 113–125.

Jansen L. A., *The Virtue in the Place: Virtue Ethics in Medicine*, "Journal of Theoretical Medicine and Bioethics" 2000, vol. 21, no. 3, p. 261–276.

Jaśtal J., *Etyka cnót, etyka charakteru*, w: *Etyka i charakter*, J. Jaśtal (ed.), Cracov 2004, s. 38.

Jost L. J., *Virtue and vice*, New York 2005; *Encyclopedia of Philosophy*, D.M. Borchert (ed.), vol. 9, Farmington Hills.

Jost L. J., *Virtue and vice*, in: *Encyclopedia of Philosophy*, D. M. Borchert (ed.), vol. 9, Farmington Hills 2005.

Kuniński M., *Komunitaryzm, czyli o bliskich związkach filozofii politycznej i socjologii*, "Diametros" 2006, vol. 3, no. 8, p. 127–131.

Machura P., *O "Dziedzictwie cnoty" Alasdaira MacIntyre'a*, "Folia Philosophica" 2002, vol. 20, no. 2, p. 71–107.

McDowell J., *The Role of Eudaimonia in Aristotle's Ethics*, in: *Essays on Aristotle's Ethics*, A. Rorty (ed.), Berkley 1980.

Murdoch I., *The Sovereignty of Good*, London 1971.

Nicholas J. L., G. Moore, *Virtue at Work: Ethics for Individuals, Managers, and Organizations*, "Philosophy of Management" 2018, vol. 17, no. 2, p. 257–259.

Parry R., *Ancient Ethical Theory*, in: *The Stanford Encyclopedia of Philosophy*, E.N. Zalta (ed.), (Fall 2014 edition), 2014, https://plato.stanford.edu/archives/fall2014/entries/ethics–ancient (Access: February 1, 2023).

Pellegrino E. D., *Towards a Virtue Based Normative Ethics for the Health Professions*, "Kennedy Institute of Ethics Journal" 1995, vol. 5, no. 3, p. 253–277.

Perret R. W., G. Pettigrove, *Hindu Virtue Ethics*, in: *The Routledge Companion to Virtue Ethics*, L. Besser-Jones, M. Slote (ed.), New York 2015, p. 51–62.

Prior W. J., *Eudaimonism and Virtue*, "The Journal of Value Inquiry" 2001, vol. 35, no. 3, p. 325–342.

Richards D. G., *Economics, Ethics, and Ancient Thought*, New York 2017.

Sandler R., *Character and Environment. A Virtue-Oriented Approach to Environmental Ethics*, New York 2007.

Sandler R., *Environmental Ethics. Theory in Practice*, New York 2018, p. 226–232).

Sim M., *Why Confucius' Ethics Is a Virtue Ethics*, in: *The Routledge Companion to Virtue Ethics*, L. Besser-Jones, M. Slote (ed.), New York 2015, p. 63–76.

Slote M., 1993, *Morals from Motives*, Oxford 2001.

Swanton C., *Virtue Ethics: a Pluralistic View*, New York 2003.

Vallor S., *Technology and the Virtues. A Philosophical Guide to a Future Worth Wanting*, Oxford 2016.

Van Zyl L., *Eudaimonistic Virtue Ethics*, in: *The Routledge Companion to Virtue Ethics*, L. Besser-Jones, M. Slote (ed.), New York 2015, p. 183–195.

Van Zyl L., *Virtue Ethics. A Contemporary Introduction*, New York 2019.

Walker R. L., *Virtue Ethics and Medicine*, in: *The Routledge Companion to Virtue Ethics*, L. Besser-Jones, M. Slote (ed.), New York 2015, p. 515–528.

Xiao Y., *Virtue Ethics as Political Philosophy: The Structure of Ethical Theory in Early Chinese Philosophy*, in: *The Routledge Companion to Virtue Ethics*, L. Besser-Jones, M. Slote (ed.), New York 2015, p. 471–489.

Zagzebski L., *Divine Motivation Theory*, New York 2004.

2 Contemporary virtue ethics in the face of the environmental crisis

Nowadays, virtue ethicists are trying to use the wealth of ancient wisdom in solving the moral problems of the modern world. One of the most important problems today is the threat to life on earth and the permanent destruction of the entire planet – the habitat of human and all other animate organisms. From nature, human beings draw natural resources, the exploitation of which is a necessary condition for human survival.[1] We are, so to speak, condemned to live on Earth because we have no other such planet (Sandler 2005, 1). Concern for the environment has been the subject of environmental ethics[2] since the 1970s, but for a long time this current referred mainly to the deontological and utilitarian tradition (cf. Treanor 2014, 8–11; Hursthouse 2016). It was only in the 1980s that there was increased interest in the approach proposed by virtue ethics; as a result, a new field of philosophical analysis was established: environmental virtue ethics.

According to Cafaro,[3] this new discipline formally began with Thomas Hill's article *Ideals of Human Excellence and Preserving Natural Environment* (1983). Hill argues that preserving the environment begins with the moral agent's inner self and motivation to act virtuously. In his article, Hill asks what kind of person would destroy the environment or even perceive its worth only in terms of cost/benefits. According to Hill, it is the character of human being, constituted by virtues, that dictates the right approach to the environment and its resources. He considers the following virtues to be the most important environmental moral dispositions: proper humility, self-acceptance, gratitude, and appreciation of the good in others. To these virtues he adds rationality, which is characteristic of a virtuous person, and he stresses that an unvirtuous person is guided by misconceptions about himself and his place in the all-world, consequently having an erroneous view of the world and making inappropriate use of resources.

Hill's article is assumed to have had the same impact on environmental virtue ethics as G.E.M. Anscombe's article had on the renaissance of virtue ethics, so both these texts are credited with symbolically launching this new discipline. However, it should be emphasized that the issue of virtues was implicitly included in environmental ethics from its very beginning.[4] Its presence can be seen even in American Transcendentalism, from which environmental ethics and environmental philosophy originated, and it is clearly visible,

DOI: 10.4324/9781003433156-4

for example, in H.D. Thoreau's iconic book *Walden, or Life in the Woods*, published in 1854. Thoreau's thought and moral example has inspired generations of environmental ethicists, philosophers, and ecologists, as well as those interested in environmental activities.

2.1 Environmental virtue ethics as a new area of philosophical reflection

The issue of environmental virtue ethics has received interest from many environmental ethicists, who see in aretology a valuable area of consideration of the human–environmental relationship. At the same time, the renewal of virtue theory and its implementation in environmental ethics does not mean a complete rejection of deontological or consequentialist ethics. Like any applied ethics, ethical reflection on the environment is subject to certain trends and follows the directions of development that are typical of the ethical discussions of its era. In the Enlightenment traditions, referring to the category of a good deed for the sake of duty or for the consequences resulting from it is still a common way of justifying moral decisions. New interpretations of representational ethics' approaches to these currents are even being developed.[5] Environmental ethics itself is very pluralistic in its nature. The content of books and articles on EVE reflects the plurality of approaches to environmental issues.

The same is true of environmental virtue ethics, which, despite being a young discipline, has already received some interesting interpretations and concepts. This chapter takes a closer look at selected theoretical problems raised within this discipline, as well as criticism of EVE and ways of responding to it.

Environmental virtue ethics has received attention from many ethicists who consider virtue reflection an important voice for environmental problems. As Cafaro (2001, 3) points out,

> Over the past twenty-five years, much scholarship in environmental ethics has focused on the intrinsic value or moral considerability of nonhuman nature. This valuable work has clearly formulated many environmentalists' intuitions that the destruction, overuse, or excessive appropriation of nature is morally wrong. It has given us plausible reasons for extending moral considerability beyond our own species and limiting our conduct accordingly. In contrast, little has been written in environmental ethics from a virtue ethics perspective that focuses on human excellence and flourishing.
>
> (Cafaro 2001)

According to Cafaro, environmental ethics cannot avoid the problem of virtue and is incomplete without EVE for two main reasons (Cafaro 2001, 4–5). First, an ethics that focuses solely on rights and responsibilities and evaluates human acts from this perspective misses an important point, namely determining what kind of life is ethically best and how a moral agent should develop in order to live an ethically best life and be the best version of himself/herself. Second, environmental virtue ethics provides positive arguments for environmental protection.

> Just as classical virtue ethics provided strong self-interested reasons for treating others with respect – reasons based on a person's concern for his own virtue and flourishing – so environmental virtue ethics can provide strong grounds for environmental protection. Above all, it can move us beyond our initial ethical response to environmental destruction – contrite self-abnegation – and toward a more positive, sustainable position of respectful dwelling in nature.
>
> (Cafaro 2001, 5)

For these reasons, in recent years we can see increased interest in the theory of environmental virtues. Among its main representatives, we should mention first of all Philip Cafaro, Louke van Wensveen, Ronald Sandler, Holmes Rolston III, Geoffrey Frash, John O'Neil, Thomas Jr. Hill, Brian Treanor, Lisa Newton, Rosalind Hursthouse, Val Plumwood, Allen Thompson, and Jason Kawall. The discipline is still developing, virtue discussion has become an important part of environmental reflections, and the circle of people address-ing this topic is constantly growing. The aforementioned thinkers contribute to reflection on environmental virtue ethics in various areas, either as authors of books on the subject or as critics of the discipline.

According to Cafaro (2010, 4), environmental virtue ethics has become an important part of the environmental discussion, and several ways of explain-ing human choices regarding the environment and environmental responsibil-ity have emerged:

- "By deploying a multifaceted set of virtue and vice concepts, it pro-vides a rich, nuanced descriptive and normative language for our rela-tionships and interactions with the natural environment (as described in Chapter 1 of Louke van Wensveen's 2000 Dirty virtues).
- By explicating the connection between human flourishing and nature, it complements duty-based and fear-based justifications for environ-mentally progressive behavior and policies.
- By providing an ecologically informed account of human flourishing that is attentive to the full range of environmental values, it offers an alternative to consumption-oriented conceptions of human flourishing.
- By articulating a positive, aspirational vision in which humans and nature flourish together, it provides an alternative conception of the human-nature relationship to those in which people are either villains and despoilers, or self-denying ascetics.

- By specifying the kind of character conducive to environmental appreciation and personal restraint, it contributes to sketching the parameters of a genuinely sustainable society: one which doesn't careen from one environmental problem to the next, seeking techno-fixes, but instead might really solve these problems.
- By focusing attention on the character states necessary for accomplishing lasting environmental improvement, it raises the salience of moral development and education to environmental ethics".

(van Wensveen's 2000)

Of particular importance in the ethics of environmental virtues are catalogues of virtues and vices that attempt to demonstrate what moral dispositions are important for the protection of the environment and for the establishment of appropriate human relations with nature. Another important area is the appeal to environmentally virtuous persons. The ethics of environmental virtues attempts to define the meaning of a so-called paradigmatic environmental character, also referred to in the literature as an environmental hero. This is a person "who helps a moral agent recognize his moral obligations to the environment and realize environmental virtues. An environmentally virtuous person influences a moral agent by showing the idea of environmental protection through his or her views and/or actions" (Dzwonkowska 2018b, 15). Environmentally virtuous people are individuals who are the first to raise certain environmental issues or proclaim pro-environmental views that are innovative for their time. Some of these people have become icons of the environmental movement, like Henry David Thoreau, Rachel Carson, Aldo Leopold, or John Muir.

The identification of a moral exemplar is based on inquiry into what an environmentally virtuous person would do (cf. Hursthouse 2016). However, it should be emphasized that normativity in virtue ethics is determined by so-called v-rules, such as "Do what is honest/charitable; do not do what is dishonest/uncharitable" (Hursthouse 1999). This approach is different from other normative ethics, but the use of these rules carries a certain difficulty as they contain concepts that are so broad that their application can be problematic (Hursthouse 1999, Dzwonkowska 2017; Duchliński et al., 2015). Therefore, to understand how to act it is necessary to observe a moral exemplar, like Aldo Leopold, Henry David Thoreau, or Rachel Carson. For example, watching people who have made significant environmental achievements can provide excellent inspiration and help us understand the nature of environmental virtues. According to Cafaro (2005, 37–39), although these protagonists preach different views, they are based on the same foundation. These views are united by five features that reveal the nature of virtue thinking:

1 "A desire to put economic life in its proper place" (Cafaro 2005), so that it becomes a source of a decent life rather than an engine of unlimited consumption. Each of the aforementioned moral exemplars emphasized the connection between unlimited consumption and the degradation of the natural

environment. Even Thoreau, who lived at a time when consumerism was just germinating, saw it as a threat to the essence of humanity and the environment.

2 "A commitment to science, combined with an appreciation of its limits" (Cafaro 2005). This demand refers to the inclusion of conservation knowledge in ethical research, which, according to Cafaro, is necessary for any EVE. However, science shouldn't be perceived as a tool with which to rule the environment, as was presupposed in Francis Bacon's postulate. Cafaro proposes a humble approach to science.

3 Nonanthropocentrism. All the aforementioned moral exemplars recognize that human beings are the perpetrator of environmental degradation; especially harmful is their perception of the world, which is narrowed to an anthropocentric perspective. Each of the mentioned moral exemplars seeks a way to go beyond the limitations posed by anthropocentrism and take nature into account in moral choices.

4 "An appreciation of the wild and support for wilderness protection" (Cafaro 2005). The idea of preserving pristine nature is one of the key themes of ecological philosophies discussed in environmental ethics. Viewed in this way, the concept of environmental protection antagonizes the world of intact nature and the world of artifacts, glorifying the former and seeking to reduce the extent of human interference with the environment.

5 "A bedrock belief that life is good: both human and nonhuman" (Cafaro 2005). Such a bicentric attitude means granting value to nonhuman entities and taking them into account in human moral choices. It should be mentioned that for each of the aforementioned environmentally virtuous individuals, love of nature was a motivation to actively protect it and to proclaim environmentally oriented views.

It must be emphasized that each of these moral exemplars from literature is of great importance for the creation of ecological culture and for active environmental protection. Thus, their attitudes and actions are often analyzed and serve as models and inspiration for future generations. Aldo Leopold, Henry David Thoreau, and Rachel Carson are icons of the environmental movement in the United States and are often pop culture icons too.[6] Nevertheless, every community has figures of merit for environmental protection who can serve as excellent inspiration for younger generations of environmentalists.

The observation of environmentally virtuous people has some drawbacks. Ronald Sandler points out that the personality patterns inherent in North American moral exemplars may differ from those of environmental heroes from other parts of the world (2007, 10). At the same time, this philosopher recognizes an imperfection of an epistemological nature that is related to determining who is an exemplary moral character. The very way of deciding who is an environmental hero is based on the belief that someone who has done a lot to protect the environment has environmental virtues, but this belief may be distorted or inadequate to reality (2007). We may believe that someone who has done a lot for active environmental protection has desirable moral dispositions, but we do not know the motives or circumstances behind his actions, and we often do not know these people and have no knowledge of their virtues. Hence, admiring someone for being active in environmental protection may not be sufficient to determine the virtues that are important in this area of activity. Despite these objections, Sandler stresses that it is not his purpose to belittle moral exemplars, since the qualities these individuals possess are respect worthy and deserving of recognition (2005, 2). Also, simply observing the actions of environmental heroes is just one way of determining environmental virtues. It should also be emphasized that actions are a better testimony to virtues than words. The aforementioned heroes have done significant things for environmental protection in their homelands; their actions authenticate their belief in the value of the natural environment and express virtuousness more fully than opulent ethical treatises as we can never be sure of virtue until it shows itself in action.

2.1.1   *The normative nature of EVE*

As previously indicated, moral exemplars are a response to the specific nature of the norms of environmental virtue ethics. A notable feature of the beginning of the renaissance of virtue ethics was attempts to demonstrate the superiority of this approach over the deontological and consequentialist approaches that dominated ethics at that time. According to Sandler (undated interview with Sandler), virtue ethics answers ethical questions better because it focuses on the character of the moral agent. As a result, it has greater potential to unleash a desire for action and commitment than ethics formulated along the lines of legal codes. Sandler also emphasizes the impossibility of humans' escape from the environment and their condemnation to the only Earth they have; he also stresses the role of environmental ethics in guiding these relations by providing specific norms of action and of character (2005, 1).

The influence of Hursthouse (2004), who emphasizes the specificity of norms established on the basis of virtue ethics, is evident here. This researcher points out that virtue ethics offers certain kinds of principles, which she calls v-rules. She states that these rules take the form of a formula: "Do what is honest/charitable; do not do what is dishonest/uncharitable" (Hursthouse 1999). Some object to this concept, saying that v-rules may not be very practical because they are too general. Nevertheless, Hursthouse counters this charge by arguing that the norms of deontological or consequentialist ethics are formulated at a similar level of generality (Hursthouse 1999, 36).

Ethics cannot focus on norms alone but should go deeper, appealing to something more fundamental than a specific catalogue of commands and prohibitions. Citing the views of Paul Ricoeur, Treanor assumes that ethics should begin with a goal, by which he means that ethics should strive to achieve the goal of human life, which is a good life. Living a good life means being a good person. This should be the starting point for establishing norms and principles. A mere catalogue of norms and principles only works in certain situations, but focusing on the moral agent changes the nature of ethics and makes it possible to refer to the purpose of human life, thus giving it a teleological character.

Also, ethics that addresses moral obligations in the issue of environmental protection cannot be limited to a catalogue of principles: it should go deeper and address the question of who the moral agent should be; it should analyze the dispositions, qualities, and habits that support pro-environmental actions. Any virtue ethics is linked to a narrative; in this sense, environmental virtue ethics should also be linked to a narrative.

Environmental virtue ethics – like any virtue ethics – is often contrasted with normative ethics, but this does not mean that norms cannot be formulated on its grounds. However, these are norms of a different kind. EVE is by no means limited to catalogs of rules, norms, and principles: it begins with the question of human character. Therefore, EVE is more individualistic and directed toward the realization of the good of a specific moral agent, which automatically excludes the possibility of creating a general catalog of norms and principles. This is, in the case of virtue ethics, a kind of advantage since, according to Sandler (2007, 1–2), normative ethics has increasingly begun to resemble legal codes; as sets of precepts and prohibitions, normative ethics proved to be ineffective and did not cover the fullness of the moral agent (for example, his intentions were ignored). Thus, virtue ethics primarily focuses not on norms and rules, and not on character norms and rules (if one can speak of such at all), but on the moral agent, on his character traits. Environmental virtue ethics, in Sandler's terms, serves to search for the so-called environmental character and what constitutes it, that is, environmental virtues. The basic ethical question – how to live? – which every human being asks himself, will not be answered by sets of norms, rules, or codes of conduct as these will never give a complete answer to such a question.

2.1.2 Summary

The renaissance of virtue ethics has inspired environmental virtue ethics. Among other things, environmental ethics has attempted to apply ancient philosophy to the moral problems of environmental degradation. Indeed, environmental issues have long been discussed by appealing to the deontological or consequentialist tradition. The fact that the virtues or vices of a moral agent can influence his behavior toward the environment contributed to the emergence of virtue discourse, the origin of which is considered to be Thomas Hill's article *Ideals of Human Excellence and Preserving Natural Environments* (1983), in which the author asks: "what sort of person would destroy the environment or even perceive its value only in terms of cost/benefit?" This question prompted ethicists to lean into the nature of the moral agent rather than the nature of the act.

2.2 Objections to environmental virtue ethics

Virtue ethics has received attention from practitioners of environmental ethics. It has produced three independent concepts, which will be presented in part two of this book. Although its representatives claim that this type of approach is much more effective in addressing ethical problems concerning human-environmental relations, some see several shortcomings in the discipline. Cafaro (2015) presents four main objections: two relate to the sphere of practical application of virtues, and two concern theoretical concepts. One objection concerning the application of virtues is an example of an internal critique whereby a representative of EVE wants to improve the discipline through his comments.

The first objection is that EVE does not provide sufficiently precise guidance regarding our actions. Of course, it is very effective in giving us information about which qualities to cultivate, but it does not tell us what to do in a specific situation. This objection relates to the issue of the normativity of virtue ethics discussed above. However, it should be noted that virtue ethics can be understood as normative ethics, i.e., it formulates norms for our behavior. These norms are somewhat different from the norms of deontological or utilitarian ethics, since the source of their normativity is virtue and vice, and the principles formulated are v-rules, which are a very useful tool in guiding the actions of a moral agent.

Developing virtues is also a way of dealing with different moral situations in a way that is just as effective as in situations where it is norms that determine what behavior to choose because virtue is a disposition that influences our actions and the way we behave. In particular, the virtue of prudence can be helpful in choosing the right action in a particular situation. As Sandler (2007, 98) argues, the guidance provided by virtues is not exclusively reserved for those who already possess moral dispositions, for the rules of virtue can be learned and can thus also be applied in concrete circumstances by those who do not possess the dispositions desired in a given situation. Such rules, however, have their limitations; for example, they do not provide a single answer to every ethical problem. Besides, the search for detailed norms is not the task of virtue ethics, and there is no need for precise codes of norms and rules. Virtue ethics in this view opposes this type of approach, seeing it as over-codified and thus bringing ethics closer to legal codes that do not capture the richness of the moral situations that the moral agent encounters.

The second objection to environmental virtue ethics is that it focuses too much on the individual moral agent and his or her pursuit of excellence and is not very causal in the area of influencing politics or wider social communities. This objection is an example of an internal critique by a representative of environmental virtue ethics, Brian Treanor. He points out that environmental virtue ethics lacks a kind of 'virtue politics' (2010, 27), which he understands as a kind of collective actions, traits, or dispositions. These are specific virtues of public life that allow the moral agent to act not only for his or her own good, but also for the good of the wider community. It is difficult to disagree with this postulate. A number of ecological studies describing the state of the environment outline the dramatic impact of harmful human activities on their natural surroundings. Many ecosystems have been severely damaged, often completely degraded precisely as a result of human actions. Hence, an ethics of environmental virtues cannot focus only on the perfection of the virtues of the individual moral agent: these virtues must also have a constructive impact on the environment in which the moral agent functions.

Several theoretical studies propose a solution to this problem. According to Cafaro (2015, 434) and Lisa Newton (2003), environmental virtues can be used to create policies that promote sustainable development principles. A similar approach can be found in Louke van Wensveen. Although this Dutch researcher does not address implementing virtues in policy, she makes the bold claim that the criterion for environmental virtue is whether it serves to achieve/sustain ecological sustainability (2000; 2001). The fulfilment of this condition is a guarantee that the individual moral agent, guided by moral dispositions, will act not only in their own interest but also for the good of the community. The very understanding of the concept of community here is broad as it is not limited to personal entities but encompasses the entire ecosystem. The community is an entity understood along the lines of the biotic community of life in Aldo Leopold's concept.

The third objection concerns not so much the practical application of environmental virtue ethics as its theory, which critics claim is insufficiently developed (cf. Cafaro 2008, 376–377). One subject of criticism is the extensive catalogue of environmental virtues and vices compiled by van Wensveen, which is criticized for not being entirely clear about how such an abundance of virtues and vices can be used. Moreover, the method of its compilation raises doubts as it is a simple list of all virtues and an analysis of their frequency of occurrence in environmental literature. Such a tactic is more reminiscent of the method used in the empirical sciences rather than the method characteristic of philosophy because it focuses on a quantitative approach rather than philosophical analyses. Van Wensveen is aware that the method used is not the most perfect one, and she somewhat ironically calls this catalogue a 'beauty contest' as it shows which virtues are most popular in ethical debate.

Methods aside, my objection is that the catalog contains terms that have not previously been referred to as virtues in the debate and do not appear to be so. One example is fear, but is this really a virtue? Van Wensveen seems to confuse emotion with virtue. Many other 'virtues' from the catalogue are similarly questionable, such as a humor, poverty (voluntary) or rage (cf. van Wensveen 2000, 163–165). The connection between these virtues and environmental protection is not obvious, and some of them cannot be defined as moral dispositions. Besides, the catalogue of virtues itself also contains words hitherto regarded as vices, such as the aforementioned rage. Additionally, adding such a term to a list that also includes gentleness creates a contradiction. The catalogue is extensive, and it is difficult to see what reasons were behind the inclusion of certain dispositions, virtues, emotions, vices or even actions in the list. Undoubtedly, the collection created by van Wensveen, although underdeveloped, has triggered an important discussion on environmental virtues. It Is clear that this Dutch writer has not defined the meaning of the term 'virtue' and uses it to describe various qualities, emotions, actions that have nothing to do with virtue.

In the discussion that followed the publication of van Wensveen's catalogue, researchers tried to describe what constitutes an ecological virtue, which she claimed is something that contributes to the well-being of the ecosystem.

According to van Wensveen, ecosystem sustainability is an indicator of the occurrence of a particular environmental virtue (2001). This condition is the so-called 'ecologically sustainable virtue' criterion. Intuitive thinking about the criterion of ecologically sustainable virtue can be strengthened by a simple syllogism proposed by van Wensveen:

1   Ecosystem sustainability is a necessary condition for the cultivation of a virtue. (…)
2   A genuine virtue includes the goal of ensuring necessary conditions for its cultivation. (…)
3   A genuine virtue includes the goal of ensuring ecosystem sustainability.

(2001, 232–233)

Since nurturing virtue requires ensuring the conditions for its development over time, which entails the preservation of the environment, authentic virtue must strive to ensure the sustainability of the ecosystem. In order for a virtue to meet the criterion of ecological sustainability, it must be 'purified' of an unecological mindset. An example of a virtue that can lead to ecosystem sustainability, once the traditionally established meaning has been changed, is genuine courage (2001, 238–239), which plays a key role in environmental protection as it constitutes the virtues of perseverance and is essential to the manifestation of ecological attitudes. The culturally perpetuated image of valor, which is also often transferred to ecology, can sanction human oppression of the environment, therefore it needs to be reformulated and re-contextualized (2001). Thus, courage is driven not by fear but by feelings of love, manifested as kindness and concern for other beings (2001, 141). However, as van Wensveen points out, this is an ideal that is not a necessary condition for the existence of virtue, and at a level attainable by average people it is sufficient for valor to be characterized by pragmatism and to be driven by vulnerability and feelings of love (2001, 140).

The criterion of environmental sustainability itself is an interesting category and an example of how environmental virtues can be determined. Nevertheless, looking at the issue from the meta-subjective side, four main strategies (Sandler 2005, 4–6) dominate in environmental virtue ethics when determining which virtue is an environmental virtue[7]. The first strategy is to extend the standard interpersonal virtue and apply it to non-human beings or even the natural environment as a whole. The second strategy refers to an appeal to the benefits of the moral agent. The third strategy refers to concern for the personal development of the moral agent. What makes a given disposition an environmental virtue also makes the moral agent a good person. The fourth strategy is related to the concept of the environmentally virtuous person who is a morally outstanding individual. Watching the moral exemplar can help identify which dispositions are environmental virtues. The development of these strategies was a response to the charge of a lack of criteria for determining which disposition is an environmental virtue.

The fourth objection to EVE concerns its anthropocentric nature. Anthropocentrism is considered to be one of the main sources of the crisis of EVE (cf. Dzwonkowska 2018a). Jaśtal even claims that "Aristotelian ethics is so extremely anthropocentric that it stands out even in the whole anthropocentric ancient tradition" (2006, 45). At the same time, he immediately adds that "this does not mean, however, that one cannot attempt to include the environmental aspect within Aristotelian virtue, although this undoubtedly requires some significant modifications to this kind of moral philosophy" (2006). I will look at the first part of Jaśtal's statement, namely the accusation that virtue ethics is anthropocentric, and where this accusation comes from. It is extremely important to see how EVE deals with the problem of the positions of human beings in the world in terms of environmental ethics and virtue ethics, because while virtue ethics is regarded as the most anthropocentric of ethics, environmental ethics is considered the least anthropocentric and often anti-anthropocentric. Ben Minteer even writes that many environmental philosophers claim that the aim of their discipline is to combat the anthropocentric approach (2008, 58). Indeed, the charge of anthropocentrism is taken very seriously by environmental philosophers: being labelled as an anthropocentrist seems to undermine the value of the solutions proposed in philosophical reflection because an anthropocentrist cannot consider the fate of other entities in a way that is honest and that ensures that every element of nature has its rightful place in our ethical choices.

The main reason why anthropocentrism has come under condemnation is that it is seen as the cause of the ecological crisis (2008). The most influential critiques of the anthropocentric position came from two texts, *Is There a Need for a New, an Environmental Ethic?* (Routley 1973), and the landmark article *The Historical Roots of Our Ecological Crisis* (White 1967). The theses articulated in these articles set the anti-anthropocentric tone for environmental discussions in the 1970s and 1980s, making criticism of anthropocentrism a cornerstone of environmental philosophy. According to Minteer (2008, 58), it was mainly L. White's article that turned environmental philosophers towards attempts to overcome anthropocentrism. His indictment of Judeo-Christian religions as a source of environmental degradation was readily accepted by much of the research community.

It seems reasonable to begin my own reflections by defining what anthropocentrism actually is as this will allow us to better understand what the anthropocentric character of virtue ethics is. The term 'anthropocentrism' itself is used in many different senses, and there are also different ways of defining it. It is worth adopting Minteer's distinction (Minteer 2008) in three types of anthropocentrism: ontological, epistemological, and ethical. Each of these relates, respectively, to a different sphere of human functioning: to one's place in the world, to the cognitive sphere, and to morality. The division into these three types of anthropocentrism is crucial for ethical discussion since each type can be overcome differently and has a different impact on the moral sphere.

Ontological anthropocentrism is a position that speaks of humans' privileged position in the world. Graphically, this can be put in the form of hierarchically ordered entities in which human beings are above the rest of the world;

this type of view is typical of neo-Thomist philosophies and is firmly established in the Christian tradition, therefore it is thought to dominate Western culture (cf. White 1967). It was the first to be criticized. This form of anthropocentrism influences the moral agent's perception of the world and their decisions regarding who should be included in the ethical debate. Enlightened anthropocentrism, however, does not necessarily imply the exclusion of non-human entities from the moral consideration, but it can retain the thesis of the privileged place of humans in the world. Thus, ontological anthropocentrism does not necessarily affect the ethical debate. The thesis that humans' privileged position in the world entitles them to exclude the natural environment from ethical choices is not supported, especially in light of the theses of so-called enlightened or weak anthropocentrism, which can take the position of ontological anthropocentrism. Thus, ontological anthropocentrism does not necessarily affect ethical judgements.

The second form of anthropocentrism is epistemological anthropocentrism, a position that is concerned with the way in which human beings come to know the world. Despite Leopold's proposals to 'think like a mountain', we only know the human way of thinking, nor will we ever know what it is like to be a bat.[8] For this reason, the objection is often made in the environmental literature that we only know the human perspective, and our thinking about the world will always be anthropocentric, so any environmental ethics must be anthropocentric. To paraphrase Holmes Rolston III (2005), this is half true, but it is dangerous if we take the whole statement to be true. Indeed, any environmental ethic is anthropocentric in the sense that it is created by human beings from their perspective. Moreover, any philosophy or ethics is anthropocentric for the same reason, but the impossibility of literally knowing the perspective of a mountain or a bat does not necessarily exclude the natural world from the realm of our moral choices. We still have the possibility to make choices in which we take the natural world into account.

This type of attitude allows us to overcome ethical anthropocentrism, which limits our moral choices to humans alone. Ethical anthropocentrism is the only position that is relevant to our moral choices and draws a line between morally significant entities and those that are not morally significant. It is based on the recognition of the value of individual entities. The non-anthropocentric stance is concerned with value that is independent of our utility, which in the literature is referred to as intrinsic or innate value. The recognition of the value of nonhuman beings is on an epistemological level, for it is human being who values particular entities, but the act of valuing itself is not the same as assigning value. Sandler points out that environmental virtues make it possible to recognize the value of entities in the environment and respond in a manner adequate to the situation. At the same time, it should be emphasized that failure to recognize the value of entities that are part of the environment does not mean that they have no intrinsic value. After all, being valued is not the same as having value.

In summary, it can be said that ethical anthropocentrism is crucial for environmental ethics. The argument that we cannot build a non-anthropocentric ethics because we see the world only from a human perspective is inadequate. We do not have to 'think like a mountain' or 'be a bat' to be able to consider the environment in our ethical choices. Even when we think only in human terms, we can make value judgments about the environment that will provide us with arguments for transcending ethical anthropocentrism. Ontological anthropocentrism also does not close humans to pro-environmental ethical decisions. One can recognize humans' ontological privileged position in the world whilst also advocating the protection of the natural environment. The only position that is relevant to ethical discussion is ethical anthropocentrism. Besides, the inclusion of virtue ethics in the ecological discussion has lagged behind the renaissance of virtue ethics precisely because of the charge of anthropocentrism. The main source of the charge of anthropocentricity of virtue ethics is its eudaimonistic nature because it was believed that an ethic focused on the pursuit of happiness of the moral agent is not an ethic that can be adequate for solving global problems, such as environmental problems. Nevertheless, eudaimonism in no way interferes with the acceptance of moral obligations to the environment.

2.2.1 Summary

Environmental virtue ethics has faced internal and external criticism. The most important objections concern the normativity of environmental virtue ethics. Since virtue ethics is based on v-rules instead of the rules offered by deontological ethics, it appears to be an ethics that is incapable of formulating norms analogous to those used in other ethical traditions. Another objection to EVE relates to its poorly developed theory – catalogs of environmental virtues have been particularly criticized. The comprehensive catalog developed by van Wensveen was questioned on the grounds that it failed to define clear criteria for deciding what an environmental virtue is. As I show in this chapter, several proposals for determining what an environmental virtue is have been put forward in response to this objection. The third objection was put forward by Brian Treanor, who believed that virtues should be applied to practice. In response to this objection, in this monograph I present Treanor's conception of virtue ethics, in which the practical dimension is an important element of EVE. The fourth objection concerns the anthropocentric nature of virtue ethics. However, as I show above, environmental virtue ethics can be as non-anthropocentric as any other account of environmental ethics. I began my analysis of this charge by sorting out the terminological confusion over the term 'anthropocentrism' in order to then move on to show that environmental virtue ethics can include both human and non-human beings in its reflection.

Notes

1 Often, the concept of the benefits humans derive from the environment is presented in the form of the concept of ecosystem services. These include provisioning services (natural resources of the environment, such as water, food or other raw materials); regulating services (related to the functions of the environment, such as regulating the climate, hydrological cycles or preventing floods or soil erosion), supporting services (which encompass the ecosystem processes necessary for other services, such as supporting biodiversity), and cultural services (which relate to values related to culture, such as aesthetic, recreational, scientific or religious ones). Although this type of approach is quite functional, especially when combined with the valuation of ecosystem services, it provides a factual argument for environmental policy. Nevertheless, it has also been criticized for sanctioning the exploitative nature of the human–environment relationship. These objections can be found in Plumwood and Peterson. The nature of this exploitative character is visible in the language, as the word 'service' is derived from the word for slave (English: *servant*, Latin: *servus*). Thus, this name, according to the cited authors, seems to be built on the servile role that the environment plays in relation to humans (Peterson 2012, 5; Plumwood 2001, 20). Moreover, it should also be mentioned that the argument regarding protecting the environment for its role in providing resources is very anthropocentric one, thus raising questions about the recognition of the intrinsic value of nature.

2 By environmental ethics, following Włodzimierz Tyburski, I mean an applied ethics that deals with humans' relationship with animate and inanimate nature. Its subject is the values and norms that can or do regulate this relationship (Cf. Tyburski 1999, 97). Environmental ethics was initiated in the 1970s. In the American literature, it is assumed that it draws its inspiration from American transcendentalism (early 19th century).

3 Cafaro in the article *Environmental Virtue Ethics. Special Issue: Introduction* (2010) writes about the importance of Hill's article as the beginning of EVE. However, regardless of the origins of environmental virtue ethics, an analysis of Cafaro's views indicates that environmental virtues discussion was already present in literature. For example, in writings of American transcendentalism's representatives – mainly Thoreau – and later also in figures actively associated with environmental protection and efforts to popularize environmental protection, such as A. Leopold and R. Carson. It should be noted that the latter are treated rather as exemplary figures who adhered to the ethics of environmental virtues in their lives (Cf. Cafaro 2008; Hill 1983; O'Neill 1993).

4 For example, van Wensveen argues that virtue language has been present in environmental discussions since the emergence of ecological reflection in ethics. More on this is discussed later in the book (cf. van Wensveen 2000, 2005).

5 As, for example, in an article whose author aims to develop Kantian moral theory regarding the environment, according to which duty and the resulting precepts and principles are the foundation of environmental ethics (Cf. Biasetti 2015).

6 Such as Thoreau, whose reconstructed cottage on Walden Pond is an important spot on the New England tourist map. His ideas of returning to nature inspired artists to create cinematic praise of life in the wild. Their work resulted in *Into The Wild* (2007).

7 A more detailed description can be found in Chapter 5.

8 A reference to an article by Thomas Nagel, who, criticizing reductionist theories that reduce the mind to corporeal states, cites the example of the impossibility of feeling what it is like to be a bat (Cf. Nagel 1974).

References

Biasetti P., *From Beauty to Love: A Kantian Way to Environmental Moral Theory?*, "Environmental Philosophy" 2015, vol. 12, no. 2, p. 139–160.

Cafaro P., *Thoreau, Leopold, and Carson: Towards an Environmental Virtue Ethics*, "Environmental Ethics" 2001, vol. 23, no. 1, p. 3–17.

Cafaro P., *Thoreau, Leopold, and Carson: Towards an Environmental Virtue Ethics*, in: *Environmental Virtue Ethics*, R. Sandler, P. Cafaro (ed.), New York 2005, p. 31–44.

Cafaro P., *Virtue Ethics*, in: *Encyclopedia of Environmental Ethics and Philosophy*, J. B. Callicott, R. Frodeman (ed.), Farmington Hills 2008, p. 375–378.

Cafaro P., *Environmental Virtue Ethics. Special Issue: Introduction*, "Journal of Agricultural and Environmental Ethics" 2010, vol. 23, no. 1, p. 3–7.

Cafaro P., *Environmental Virtue Ethics*, in: *The Routledge Companion to Virtue Ethics*, L. Besser-Jones, M. Slote (ed.), New York 2015, p. 427–444.

Duchliński P., Kobyliński A., Moń R., Podrez E., *O normatywności w etyce*, Cracow 2015.

Dzwonkowska D., *Normatywność etyki cnót środowiskowych na przykładzie etyki Ronalda Sandlera. Komentarz*, "Avant" 2017, vol. 8, no. 3, p. 99–108.

Dzwonkowska D., *Is Environmental Virtue Ethics Anthropocentric?*, "Journal of Agricultural and Environmental Ethics" 2018a, vol. 31, no. 6, 723–738.

Dzwonkowska D., *Jan Gwalbert Pawlikowski jako przykład wzorcowego charakteru środowiskowego*, "Etyka" 2018b, vol. 56, p. 13–28.

Hill T., *Ideals of Human Excellence and Preserving Natural Environment*, "Environmental Ethics" 1983, vol. 5, no. 3, p. 211–224.

Hursthouse R., *On Virtue Ethics*, Oxford 1999.

Hursthouse R., *Normatywna etyka cnót*, in: *Etyka i haracter*, J. Jaśtal (ed.), Cracov 2004, p. 191–209.

Hursthouse R., *Virtue Ethics*, in: *Stanford Encyclopedia of Philosophy*, E.N. Zalta (ed.) (Winter 2016 edition), 2016, https://plato.stanford.edu/archives/win2016/entries/ethics-virtue (Access: July 15, 2023).

Jaśtal J., *Etyka cnót wobec wyzwań etyki środowiskowej: spór o granice naturalistycznego dyskursu etycznego*, "Diametros" 2006, vol. 3, no. 9, p. 34–50.

Minteer B. A., *Anthropocentrism*, in: *Encyclopedia of Environmental Ethics and Philosophy*, J. B. Callicott, R. Frodeman (ed.), Farmington Hills 2008, p. 58–62.

Nagel T., *What Is It Like to Be a Bat?*, "The Philosophical Review" 1974, vol. 83, no. 4, p. 435–450.

Newton L., *Ethics and Sustainability: Sustainable Development and the Moral Life*, New Jersey 2003.

O'Neill J., Ecology, *Policy, and Politics: Human Well-Being and the Natural World*, London–New York 1993.

Peterson K., *Ecosystem Services, Nonhuman Agencies, and Diffuse Dependence*, "Environmental Philosophy" 2012, vol. 9, no. 2, p. 1–19.

Plumwood V., *Nature as Agency and the Prospects for a Progressive Naturalism*, "Capitalism, Nature, Socialism" 2001, vol. 12, no. 4, p. 3–32.

Rolston III H., *Environmental Virtue Ethics: Half the Truth but Dangerous as a Whole*, in: *Environmental Virtue Ethics*, R. Sandler, P. Cafaro (ed.), Oxford 2005, p. 61–78.

Routley R., *Is There a Need for a New, an Environmental Ethic?*, "*Proceedings, Twelfth World Congress of Philosophy*" 1973, p. 205–210.

Sandler R., *Introduction: Environmental Virtue Ethics*, in: *Environmental Virtue Ethics*, R. Sandler, P. Cafaro (ed.), Oxford 2005.

Sandler R., *Character and Environment. A Virtue-Oriented Approach to Environmental Ethics*, New York 2007.

Sandler R., *An Interview with Ronald Sandler*, https://cup.columbia.edu/author-interviews/sandler-character-environment (Access: July 31, 2023).

Treanor B., *Environmentalism and Public Virtue*, in: *Virtue Ethics and the Environment*, P. Cafaro, R. Sandler (ed.), Dordrecht 2010, p. 9–28.

Treanor B., *Emplotting Virtue: A Narrative Approach to Environmental Virtue Ethics*, Albany 2014.

Tyburski W., *Główne kierunki i zasady etyki środowiskowej*, in: *Wprowadzenie do filozoficznych problemów ekologii*, A. Papuziński (ed.), Bydgoszcz 1999, p. 97–132.

Van Wensveen L., *Dirty Virtues: The Emergence of Ecological Virtue Ethics*, Amherst 2000.

Van Wensveen L., *Ecosystem Sustainability as a Criterion for Genuine Virtue*, "Environmental Ethics" 2001, vol. 23, no. 3, p. 227–241.

Van Wensveen L., The Emergence of Ecological Virtue Language, in: *Environmental Virtue Ethics*, *Environmental Virtue Ethics*, R. Sandler, P. Cafaro (ed.), Oxford 2005, p. 15–30.

White L., *The Historical Roots of Our Ecologic Crisis*, "Science" 1967, vol. 155, no. 3767, p. 1203–1207.

3   The language of virtue ethics

The word 'virtue' is somewhat troublesome as "it carries the stigma of sounding old-fashioned, preachy and self-righteous" (van Wensveen 2000, 6). Nevertheless, it is one of the most important concepts in ethics, and nowadays – thanks to the works of Anscombe (1958) and MacIntyre (1985) – the topic of virtues is increasingly the subject of ethical discussions. In this monograph I put forward the thesis that the ethics of environmental virtues must take virtues into consideration. Linking virtue to the sphere of action, however, requires that the practical dimension of morality is reflected in the language of virtues, or more precisely, in philosophical concepts pertaining to the phenomenon discussed in ethical theory. Hence, in this monograph I devote considerable attention to the language of virtues. MacIntyre emphasizes that moral concepts change along with changes in social life (1985, 29). By tracing the transformations taking place in the meaning of the terms 'virtue' and 'vice,' it is possible to understand what context has been given to these words.

3.1 Preliminary terminological distinctions

3.1.1 'Arete' and translation problems

The Greek term ἀρετή means "goodness, accuracy, splendor, power, proficiency, beauty, honor, happiness, prosperity, fertility, moral goodness, greatness of soul, virtue, reliability, magnanimity (…) service, merit, innocence, dexterity; valor" (Węclewski 1884, 102); in the plural, it means "heroic deeds; fame, heroic glory" (Jaeger 1962, 37). The Greek word ἀρετή captures the ideal of "nobility, moral virtue, excellence (…) literally (means) 'that which is good'" (Online etymology dictionary, 2022). The higher degree of *arete* is *areion*, and the highest degree is *aristos*, from which the word 'aristocracy' is derived. As MacIntyre notes, according to Aristotle virtues are available for the noble-born that are beyond the reach of others. Virtues of people who enjoy wealth and high social status are virtues that are beyond the reach of a poor man, even if he is free (cf. 1985). This applies to the virtues most important in human life, such as magnanimity and generosity.

DOI: 10.4324/9781003433156-5

Besides, the term 'virtue' always meant the best in a noble person, and in Homer's time it meant not only the virtues of men but also the power of the gods or the qualities of racehorses (cf. Jaeger 1962, 426). The common man does not possess ἀρετή because Zeus deprived him of half of his virtues; moreover, as the etymology of the words 'virtue' and 'aristocracy' demonstrates, virtue is only afforded to the noble-born. "In the eyes of the Greeks, exceptional achievement and ability have always been an indisputable condition for occupying a presiding position. *Arete* and dominion are inextricably linked" (Jaeger 1962, 38). *Arete* is for Homer mainly the strength and dexterity of a warrior – the term 'virtue' was not used in a moral sense at that time. In ancient times, virtue meant the virtue of heroic valor, which is a quality of those who stand up for themselves and engage in combat. Jaeger, however, stresses that it is unlikely that the term was narrowed down in the vernacular exclusively to the meaning in which it was used by Homer. *Arete* as heroic strength and valor is preserved in the language of songs that praise heroes. Thus understood, *arete* was connected with the function that a man performs in the community because his main duty in Homer's times was to protect society and to serve as a warrior. His bravery is his visiting card. Moreover, this brave warrior, fulfilling heroic deeds, wants to gain recognition for his achievements, for the virtue of valor and strength, which were so valued by ancient culture.

Valor, strength, and bravery's association with the desirable qualities of a man is not in doubt due to the cultural context of a society in which men were constantly forced to go into battle.

> The link between the concepts of nobility and chivalrous bravado is also evident in the adjective *agathos*, which is close to the noun *arete*, though formed from a different base, and which already means 'nobly born', 'brave' or 'efficient', while it does not yet show its later meaning of 'good' (in general, like *arete*, it does not yet mean moral 'virtue').
>
> (Jaeger 1962, 38–39)

Being of noble birth implies higher social demands on those who are, by definition, supposedly closer to the ideal than common people. Noble ancestors and their achievements compel the aristocracy to persist in virtue, and at the same time noble qualities are closely linked to the possession of virtue. Thus, the *aristoi* (noble man, aristocrat) compete in the possession of *arete*. Here again, the appreciation comes to the fore that the *aristoi* possess *arete* due to their lineage.

An additional difficulty in understanding the concept of virtue is that the Greek term *arete* underwent changes in meaning.

> For it meant, in its original meaning, valor, heroism; later (much broader) it meant the adaptation of individual units and of each thing to fulfill its proper peculiar task, (...); finally, its meaning became narrower and meant the moral qualities of a man, his virtuousness and his virtues (Hellenistic. e.g., Stoic and Christian times).
>
> (Gromska 2007, 53)

The changes in the meaning of the word *arete* can also be traced in philosophical texts. For example, in *Meno*, Plato writes that

> the virtue of a man – he should know how to administer the state, and in the administration of it to benefit his friends and harm his enemies; and he must also be careful not to suffer harm himself. A woman's virtue, if you wish to know about that, may also be easily described: her duty is to order her house, and keep what is indoors, and obey her husband. Every age, every condition of life, young or old, male, or female, slave or free, has a different virtue: there are countless virtues and no lack of definitions of them; for virtue is relative to the actions and ages of each of us in all that we do. And the same may be said of vice.
>
> (Plato n.d., 2)

Virtue, therefore, originally meant proficiency in daily activities and efficiency in performing daily duties. In the ancient world, where rights were vested in gods and free citizens, the most important virtues were courage, moderation, and justice (cf. Parry 2014). Justice was seen as a moral disposition to ensure that everyone received what was due to him. Courage was a disposition to act diligently to achieve the right goals in every situation, while temperance was a disposition that helps one deal equally with appetite and emotions. Such a broad view of virtue goes beyond morality and allows virtue to manifest itself in many spheres of human life. Tatarkiewicz emphasizes that, according to Aristotle

> there are as many virtues as there are activities proper to man, because every activity has its virtue; when, for example, a man is dealing with external goods, then generosity is a virtue; when, again, he is overcome by a feeling of fear, then valor is a virtue.
>
> (2014, 89)

The richness of meaning of the term ἀρετή leads to many difficulties in terms of translation to national languages. Also, the Latin *virtus* is ambiguous and saturated in content, which poses translation difficulties; the word 'virtue' is derived from Latin, while in romance languages the terms are *vertu* (French) and or *virtù* (Italian). The problem is that these terms are ambiguous, just like their Latin prototype. Etymologically, the term *virtus* is derived from the words *vis* and *vir*, meaning 'strength' and 'man,' respectively, and this word is traditionally associated with valor. Thus, some older dictionaries define it as "whatever constitutes manliness as to sex, courage, boldness, etc., in comparison with the female sex" (Woelke 1831, 785–6), or as a second meaning, "valor, bravery" (Woelke 1831, 786).

Virtue, therefore, primarily meant valor, bravery, heroism, and doing one's duty to protect the state. Over time, however, "technical inventions have slowly contributed to the devaluation of men's valor. The invention of gunpowder in the 16th century gradually changed the technique of combat, in which enemies had hitherto struggled breast to breast" (Ossowska 1992, 41). Hence, the meaning of the term also carries other connotations as it is no longer only associated with strength (Latin: *vis*) or with a man (Latin: *vir*).

Virtue, however, is not a term exclusively associated with a human being, as Gromska points out:

> the Romans, however, also spoke of the virtue of horses (Cicero, Leges 1, 16), herbs (Ovid., Met. XIV 356), trees (Cicero, l.c.), ships (Liv. 37, 24), wine (Propert, 3,15, 20), iron (Iustin., 11, 14 fin.), parts of speech (Quintil., 4,2), and pronunciation (Tenn., 1,5; 8 praef.)
>
> (Gromska 2007, 59)

The meaning of *virtus* as virtue was thus not unique; the Latin term is much more capacious, and narrowing it down to mean only 'virtue' impoverishes its content. Virtue is also

> bravery, efficacy, aid *deorum equi*: hence (hence – D.D.) a miracle
>
> (Woelke 1831, 785–786)

but above all it is "virtuous conduct, virtue, benevolence, *est tauta virtute*, (…) good, special attribute" (Woelke 1831). Virtue has not only a moral dimension but can also be a virtue of the intellect – a virtue related to cognition, not just to ethics. As one of the meanings of the term *virtus*, the Latin–Polish dictionary defines it as a set of virtues:

> a. intellectual: skill, especially entitling one to perform some function, competence; b. moral: high ethical standard, virtue; c. (of things) good quality"
>
> (Plezia 1999, 630)

This multiplicity of meanings of the Greek and Latin equivalents of the term *virtus* has made it difficult for philologists and philosophers to translate them each to a single word.

"In Book II, Chapter 5 of *Nicomachean Ethics*, Aristotle famously identifies virtues as *hexeis* (sing. *hexis*)" (Lu 2015, 197). The term ἕξις is translated as 'habit,' 'disposition,' 'state,' 'active condition.' Thus, there arises a question regarding what, if any, is the difference between *arete and hexis*. Mathew T. Lu (2015, 197) claims that the dynamic element of *hexis* is central to properly grasping the meaning of this term. He emphasizes that *hexeis* in some specific way links the active and passive elements and has a causal function. It is the *hexeis* in the soul of the virtuous person that contributes to him acting properly. According to Lu, when a virtuous person acts, he acts on account of *hexis* (2015, 203). Lu argues that although normative ethics focuses more on character than on action, *hexis* understood properly does in fact have an active agent. The focus is on action – more specifically on the action that a virtuous person performs. According to Lu, virtues, including intellectual virtues, are *hexeis*, and this requires a rethinking and deeper understanding of practical wisdom. Without this, not only will modern virtue ethics not be a 'valuable addition' but will be contrary to the idea of virtue ethics as conceived by Aristotle.

3.1.2 The opposite of virtue

Virtue as fitness has its negative counterpart in vice, which is its opposite. The term 'vice' has far fewer meanings: it is understood as "a negative quality, detrimental to one's value, excellence, vice, deficiency, imperfection, defect; deviation, immorality, foulness, defect" (Karłowicz et al. 1953, 441). The Latin term *vitium* means "defect, imperfection, disadvantage, vice, deficiency, deficiency, defect" (Korpanty 2003, 982). A defect can mean an imperfection of both body and spirit – a moral defect as well as a bodily defect, disorder or ailment. It can also be an unfavorable omen, an unfavorable sign. The verb with the same root, *vitio*, means "to spoil something, to disfigure, to pollute, to stain" (Korpanty 2003), and figuratively it means "to adulterate, to distort; for religious reasons to declare inauspicious, unsuitable for public activity (*auspicium diem*); to poison" (Korpanty 2003) The adjective *vitiosus*, derived from *vitium*, means "defective, flawed; imperfect, inferior; morally bad, wicked," and in the plural it means "calamity, misfortune" (Węclewski 1884, 366).

The Greek κακια means malice, ineptitude, infirmity, cowardice; immorality, corruption, recklessness, malice, criminality; bad name, infamy (Liddell et al., 1940), but also bad reputation. In the pantheon of deities, *Kakia* is the personification of the spirit (daimon) of vice and immorality. She was depicted as a vain, plump woman wearing a garment that revealed her charms. In mythology, she appears before Heracles together with Arete when he is pondering which path to take in life: that of virtue or that of immorality (*kakia*). From Arete, dressed in a white dress, with eyes filled with modesty, there is even a glow (*Kakia*, n.d.). The second woman, Kakia, was plump and soft, with her face painted to enhance the natural whites and pinks, and she dressed to emphasize her charms.

She tempted Heracles, promising that if he chose her, he would follow the most pleasant and easy path, taste all the sweet things of life and not know the hardships. Instead of thinking about wars and worries, he would only consider what drink or food to choose, what sight or sound would please him the most, what touch or perfume – which love could give him the most joy, which bed is the most comfortable. Kakia promised to obtain all this and even more effortlessly, through the fruits of other people's labor. When Heracles asked for her name, she replied that she is called Happiness by her friends and Kakia by those who hate her.

Arete, too, tried to attract Heracles to her path, expressing the hope that he will not be deceived by pleasant words, since everything good in the world requires effort. Arete claims that Heracles's good upbringing will make him choose the path of virtue and noble deeds. She trusts that he will make efforts to achieve good. However, this requires sacrifice. If he wants the love of his friends, he must do good toward them; if he wants honors from the city, he must help the city; if he wants to reap an abundant harvest, he must cultivate the land; if he wants the profits of the sheepfold, he must care for it; if he wants to gain glory through war, he must learn the art of war and know how to use it properly; and if he wants the body to be strong, he must accustom it to be the servant of the mind and train it by putting in the effort.

Kakia did not give up, tempting Heracles with her beauty and promises of profit and effortless benefits, while Arete assured him that, although effort was needed to achieve goodness, it brought long-lasting rather than fleeting benefits. Thus, Prodicus of Ceos (5th century BC) describes how virtue and vice sought to seduce a young man entering adulthood – how both tried to convince him that their path was the better life choice for a young person. In the representations of this scene, one can see a significant difference between the two life paths (*Kakia*, n.d.). On the one hand, we have Kakia, adorned with gold and necklaces, in a purple robe, with painted cheeks and eyes highlighted with henna, with braided hair and golden slippers. Her figure not only tempts with the promise of an easy and pleasant life, but she herself shows by her superficiality that she lacks nothing. Arete, on the other hand, gives the impression of a tired woman, looking miserable and wretched, for she has come barefoot in the most modest of robes. Not only is her image uninviting, but the path she offers is full of effort and hardship in place of the easy profits with which Kakia tempts. This recognition of the value of Arete is the essence of following the path of virtue, which means acting virtuously, preceded by a recognition of what is right and what is wrong.

3.1.3 Summary

I have devoted a relatively large amount of space to a linguistic consideration of virtue, mainly because I share MacIntyre's claim regarding the importance of moral concepts in society. At the same time, I believe that transformations in the use of the colloquial term also reflect cultural changes. As can be seen from the preceding description, the term has changed connotations, oscillating between the idea of valor, bravery, strength, ideal physical qualities, and the meaning given to it in ethics. In everyday language, no one associates virtue with bravery or courage anymore. Even the discussion among classical philologists and philosophers about defining 'virtue' as bravery is not reflected in the colloquial use of the term. This change reveals a cultural phenomenon that could undoubtedly be an interesting contribution to further research.

3.2 Postulate of a return[1] to the language of virtues

I base this subsection on a call for a return to the language of virtues, based on analysis of Louke van Wensveen.[2] Her book *Dirty Virtues: the Emergence of Ecological Virtue Ethics* (2000) is the first publication devoted exclusively to environmental virtues.[3] Van Wensveen's reflections are based on sociological work[4] and the study of religion (mainly the thought of Thomas Berry[5] and Thomism, interpreted in the spirit of feminism[6]), but van Wensveen also repeatedly emphasizes that she is inspired by the philosophy of Aristotle or Thomas Aquinas.[7] Although this book on dirty virtues does not present a systematic conception of environmental virtue ethics, it has played an important role in directing the attention of environmental ethicists to the issue of virtue.

3.2.1 A voice for the restoration of virtue language in environmental ethics

The main aim of van Wensveen's work is to restore the language of virtues, which the author considers to be much richer and more diverse than the language of deontological ethics, which is similar in style to legal discussions. By restoring the language of virtues, van Wensveen understands the use of terminology that defines moral depositions as environmental virtues and vices. In her view, the term 'virtue' has been replaced in environmental ethics by other terms. The terms 'attitude,' 'habit,' or 'practice' have taken over the richness of meaning of aretological terminology (Van Wensveen 2000, 7). The aim of van Wensveen's monograph is therefore to re-establish the virtue discourse and to give virtue its rightful place in ecological discussions. According to this researcher, this is crucial because the language of virtue is currently in crisis and the contemporary state of the environment prompts the adoption of an aretological narrative (Van Wensveen 2000):

> since ecologically minded people tend to perceive the current environmental crisis as extremely serious, it should not be surprising that they resort to this traditional linguistic construction that implies active engagement, even if they do not call this language by its traditional name.
> (Van Wensveen 2000, 7–8)

In *Dirty Virtues: The Emergence of Ecological Virtue Ethics*, the diversity of virtue language is presented. As the Dutch philosopher points out, her research into the language of virtues cured her of the belief that she could construct a unified theory of environmental virtues. The richness and diversity of virtue language makes this impossible (Van Wensveen 2000, 22). The environmental virtues in this conception have been described as 'dirty' for two reasons: First, they relate to working with soil, which always makes us physically dirty. As van Wensveen writes, "dirty = soil" (Van Wensveen 2000, 4). The second reason is that they are not particularly laudable virtues. Louke van Wensveen refers to the meaning of the word 'dirty' in English, which also means something indecent, unclean, and refers to some form of taboo. According to this Dutch researcher, environmental virtues are a kind of taboo.

The term 'virtue' itself is troublesome, bearing the stigma of something preachy and old-fashioned, unfashionable or even associated with only one sphere of human life (Van Wensveen 2000, 3; cf. Taliaferro 2005, 159).

As Natasza Szutta points out, in the Polish language the word 'virtue' has been ridiculed due to its persistent connotation of sexual innocence (2015, 172). This stigma is why the question of virtues and vices in the language of morality is neglected. Brian Treanor also points out the awkwardness of this notion, but he looks for the root of the problem elsewhere, arguing that the translation of the Greek word arete as 'virtu' is not the most favorable. The word 'virtue' itself is associated with Victorian prudishness and zealous piety rather than with the moral prowess of a good person (2014, 28). Because of such connotations, referring to aretology was not popular in philosophy for quite a long time, and in a sense this part of the ethical debate was associated with theology. Hence, in spite of the aforementioned short-comings, van Wensveen's work has played an important role by drawing attention to aretological issues in the discussion of the moral dimension of human–environment relations.

A return to virtue language is one of the slogans raised by contemporary virtue ethicists, who postulate a turn away from a language of ethics con-structed along the lines of law. Virtue language itself appeared in the debates of many thinkers at the turn of the previous century. This change in language reflects a transformation in the way philosophy is practiced. At the same time, aretological language permeates various philosophical disciplines (e.g., episte-mology) and public debates, and it is even sometimes used by proponents of the deontological (Onora O'Neil, Otfried Höffe) or utilitarian traditions (Julia Driver, Peter Railton) (cf. Szutta, 2015, 171). Thanks to van Wensveen, it has been increasingly heard in discussions of environmental ethics since the begin-ning of the century.

Virtue, according to van Wensveen (2000, 8), was in the environmental debate before Lynn White's iconic article *The Historical Roots of Our Ecological Crisis* (1967). This article blamed Judeo-Christian religions for the ecological crisis, thus also excluding the word 'virtue' from environmental debates. Indeed, White's views were readily adopted by many thinkers and developed into a critique of religion, therefore the notion of 'virtue,' which is associated with the theological dimension, had no basis in philosophy. Consequently, the ecologi-cal discussion was directed toward the topic of attitudes toward the environ-ment. Van Wensveen argues that by introducing the terminology of 'attitudes,' White made them substitutes for the words 'virtue' and 'vice.'

However, as van Wensveen points out, the language of virtues corresponds to efforts to live according to ecological beliefs. In her view, nurturing virtues means integrating emotions, thoughts, and actions in such a way as to form a perfectly coherent personality in which ecological ideas can best materialize. They must flow from the inner conviction of those realizing them and not be imposed by force. They represent the individual development of moral facul-ties; hence, virtue ethics is more suited to fostering moral attitudes than deon-tological ethics, which is a form of external coercion.

The possibility of applying virtue ethics to address the environmental crisis has been recognized in environmental ethics. Those involved in environmental protection speak of the importance of respect, care for nature, humility, simplicity, frugality, straightforwardness, and a range of other moral dispositions (van Wensveen 2000, 227). In addition to virtues, environmental vices such as arrogance, cruelty, consumerism, and thoughtlessness are sometimes discussed in debates. This interest in typologies of vices and virtues is evident in the various concepts of environmental virtue ethics or in the analysis of issues related to virtue theory in the environmental debate. Thanks to this lively debate, virtue ethicists do not have to start from scratch but can join the current discussion and build on existing analyses. The language of virtue has another advantage over other types of discourses:

> virtue language has premodern roots, and although it comes to us sifted through the mazes of modernity, its internal consistency and comprehensibility are not dependent on the worldview that came to reign with the scientific and industrial revolutions. Given that many critics see the modern worldview as an important factor in bringing about the ecological crisis, it will be helpful to have access to a form of moral discourse that is not too much in cahoots with this worldview.
>
> (Van Wensveen 2000, 18)

3.2.2 Features of environmental virtue discourse

The virtue discourse, according to van Wensveen, has five distinctive features (Van Wensveen 2000, 9–18). First, it is an integral discourse characterized by coherence and logic. In a sense, it is related to other ethical discourses, especially discussions of virtue ethics. Van Wensveen claims that the

> best way to understand ecological virtue language as an integral discourse is to see it as analogous to a bioregion, which has an integrity and unique character, but which connects with other bioregions and participates in the larger cycles of the biosphere.
>
> (Van Wensveen 2000, 9)

The integrity of ecological discourse stems from a practical aspect represented by environmental activism. Thus, even though there might be some disagreements in theoretical discussion, there is some sort of "irenic interpretation of the practical" (Van Wensveen 2000, 10) aspect.

The second feature of the discussion of virtue/vice issues is its diversity. Each participant in the debate perceives the key virtues differently and may also establish a different hierarchy of the most important virtues. This is due not only to the variety of ethical situations, but also to cultural and worldview differences. This also shows that the language of virtue ethics is alive, is still subject to change, and poses new ethical challenges to researchers.

The third feature of virtue discussion is its dialectical character, which means certain patterns of logical and psychological tensions can be seen in it.

"To the extent that these tensions are deliberate and produce desirable results that cannot be achieved otherwise, the discourse of ecological virtues can be called a dialectical discourse" (Van Wensveen 2000, 14). At the same time, van Wensveen stresses that not all tensions are deliberate; rather, some are contradictions typical of ethics in its early stages of development (Van Wensveen 2000). This researcher cites love as an example: she sees a dialectical contradiction in the fact that respect is needed for love to appear, while at the same time respect is defined by her as the basis of love. Her comments relate to the following words of Macy (1989, 209):

> Love is respect's strongest foundation, although it is often difficult to uncover because of our desires and fears. Nevertheless, seeing value in another, regardless of his/her/its utility to us, is a crucial first step towards discovering that love. Without respect we cannot love.

However, inconsistencies like this one will be clarified with a growth of moral reflection on our relation to the environment.

The fourth feature of virtue language is its dynamic character, as "ecowriters appreciate the ever-changing and multifaceted structure of the world" (Van Wensveen 2000, 16). The language of environmental virtues still leaves much room for development, greater precision, or refinement of its terminology. It is a language that expresses the dynamism of this young discipline, while at the same time revealing a space open to refinement of theory.

The fifth feature of virtue language is that it is a visionary discourse without a social ethics. One of the hallmarks of environmental virtue ethics is its focus on moral character, but not in relation to the better functioning of the individual in the world in terms of social relations.[8] It emphasizes the need for human involvement in caring for the well-being of the entire planet and the need to remodel social structures to best serve the protection of all forms of life on Earth. "Ecological virtue discourse thus derives much of its impetus from a social idea. Ironically, however, most ecological virtue language does not display the features of social ethics" (Van Wensveen 2000, 17). Van Wensveen's thesis about the omission of social issues is highly debatable, as most environmental virtue ethicists include the social aspect in their analyses. This is reflected in the language. In general, representatives of the discipline use the term 'environmental' rather than 'ecological' for the terms 'virtue,' 'vice' or 'ethics,' which is a declaration that they take social issues into account in their views. Nor is it true that the social and ecological dimensions appear as two separate areas in ethics, as van Wensveen claims. Environmental ethicists mostly declare a view of man as a biological-cultural being whose space of functioning is both the natural environment and the social space. Thus, the place of realization of moral dispositions is both spaces, and virtues have both social and environmental dimensions.

According to van Wensveen, virtue language is materially modern and formally traditional (Van Wensveen 2000, 107) in order to point out another feature of the discourse: the dimension of thinking in terms relating to gender. Indeed, van Wensveen believes that the formal dimension of virtues is related to thinking in terms of feminist discourse. This type of thinking speaks of an aspect of tradition that should be rejected as oppressive toward women. According to feminist theses, the structure of language is dominated by socio-cultural gender identity, which is related to the universality of virtues, which, however, refer to one gender in social practice. At the root of the division of virtues and vices into masculine and feminine lies the view long held in Western societies that women are unable to develop virtues that are particularly valued (mainly by men) (Van Wensveen 2000, 117). This fact gave rise to the need to also create moral standards for women that would be associated with their social function. Moreover, the domination of men and the lack of women's rights – a state of affairs that persisted for a long time – resulted in women being considered incapable of adhering to certain virtues, instead being more likely to manifest vices (Van Wensveen 2000, 116). Van Wensveen puts forward the thesis that the language of environmental virtues is an interesting phenomenon in this regard as many of the key environmental virtues are virtues traditionally identified with women. For example, love, care, compassion, gentleness, humility, intuition, sensitivity, openness, and the ability to cooperate are considered such virtues (Van Wensveen 2000, 118–119).

3.2.3 The richness of the language of environmental virtues

Speaking of the language of environmental virtues, it is worth looking at the catalog that van Wensveen proposes. It is a rather eclectic list of 189 environmental virtues and 174 environmental vices[9] (2000, 163–167), whose author drew inspiration from an analysis of environmental literature published from the 1970s to about the mid-1990s.[10] Van Wensveen herself retains a certain cognitive skepticism about her list of ecological virtues and vices and treats its preparation as a preliminary stage for further philosophical analysis. Besides, she says, the catalog was intended to be used to define the cardinal virtues for the ecological era, and its creation was guided by the conviction that if a certain virtue appears frequently in the literature, this means that it is of dominant importance in times of countering the ecological crisis (2005, 175). Van Wensveen's catalog, however, showed the richness of the ongoing discourse. For many philosophers, this catalog is a valuable source of information.

This Dutch philosopher describes the list as strange; in her opinion, the top part of the list is quite intuitive, but the bottom part is not. The virtues at the bottom of the list are, in her opinion, of lesser importance. The catalog itself is somewhat jokingly described by her as a "popularity contest." "After all, quantity has never been a good measure of quality" (Van Wensveen 2000). The following table shows an abbreviated version of this list of ecological virtues.

Table 3.1 Frequency of virtue terms found in a review of post-1970 environmental literature

Rank	Virtue	Frequency	Sources
The Winner	Care	79	17
Runner-up	Respect	65	12
Third place	Love	54	12
Fourth place	Compassion	34	12
	Reverence	29	12
	Humility	34	9
	Creativity	33	10
	Hope	29	9
	Sensitivity	29	6
Runners-up	Identification (with nature)	21	8
	Acceptance (of limitations)	20	9
	170 other virtues		
Barely in view	Diligence, efficiency, endurance, forgiveness, gentleness, humor, sincerity, tolerance		

Source: L. van Wensveen, *Cardinal Environmental Virtues: A Neurobiological Perspective*, in: *Environmental Virtue Ethics*, R. Sandler, P. Cafaro (ed.), New York 2005, p. 175.

Louke van Wensveen recognized that such a diverse catalog of virtues captures the full richness of human moral choices, but at the same time these nearly 200 environmental virtues can be reduced to four main groups (Van Wensveen 2000, 176–177):

1 Virtues of position – understood by van Wensveen as the constructive habit of seeing oneself in a certain place, in a relational structure, and acting in accordance with that position. "Environmental ethicists commonly argue that an ecological way of being and acting rests on seeing ourselves as responsive nodes in a complex network, rather than overbearing top dogs in a linear hierarchy" (Van Wensveen 2000, 176). This way of thinking about a moral agent's network of relationships is like Aldo Leopold's view of reality, which emphasizes that all entities are part of a biotic community. In many traditional concepts, which are criticized for excessive anthropocentrism, humans are placed at the top of this hierarchy. However, Leopold's thought is closer to holism as it equates the value of humans and other entities in the environment. The network of relationships forces active participation in the community of which one is a part, and it prompts one to respond to the needs of other members of that community and modify one's own actions. These virtues require, first of all, sensitivity to the environment of the moral agent.

As a contemporary example of virtues of position, van Wensveen cites Thomas Hill, who writes that human beings do not seem to understand that it is merely "a speck on the cosmic scene, a brief stage in the evolutionary process, only one along millions of species on Earth, and an episode in the course of human history" (1983, 217). Humility forces a departure from the pre-Copernican era in which man viewed the Earth as the center of the universe. Thus, Hill suggests virtues of proper humility that allow moral agents to recognize the proper status of the animate and inanimate entities of the natural world. According to Hill, the discovery of proper humility is the path to determining one's true position in the universe. Louke van Wensveen cites other examples of people who appreciate the importance of the virtue of position: Bill Shaw (1997), who analyzes the virtues of respect, prudence, and practical judgment in Aldo Leopold's Earth ethics; and Lisa Gerber, who emphasizes how humility helps us move from self-centeredness to a focus on the greater whole (1999). Van Wensveen also sees similar views from eco-theologians with a strong biological background (she refers to Susan Power Bratton and Celia Deane-Drummond) who allude to biblical wisdom; one can also see these virtues in texts on vulnerability (cf. Rodman 1976, Rolston 1988).

2   Van Wensveen defines the virtues of care as the habit of constructively engaging with the relational structure we are in. This means recognizing and responding to the needs of those around us and being able to engage with the entities around us. According to Geoffrey Frasz (2001), humility alone is not enough to protect our natural surroundings; friendship with the natural world is crucial in his view. It is friendship that helps us see the needs of the natural environment. Lisa Gerber (1999) lists caring and attentiveness as key values. Jennifer Welchman points out that benevolence plays a primordial role in developing an environmentally sensitive character (1999). James Nash (1991), on the other hand, takes the position that loving nature is a key value.

3   The virtues of attunement refer to the degree of constructive engagement in the environment and the habit of coping with temptations and tuning our drives and emotions to fit where we are. Tuning in "is key because without such personal adjustment all our humility and respect, our wisdom and sensitivity, our attentiveness and friendship may still amount to nothing" (van Wensveen 2005, 177).

According to van Wensveen, an example of virtue in this group is the virtue of frugality in an interpretation proposed by James A. Nash (1997) or the virtue of simplicity as seen by Philip Cafaro (1998; cf. Gambrel, Cafaro 2010) and Lisa Newton (2003). The value of simplicity, limitation, or moderation is recognized by many thinkers and often appears as part of a critique of consumer culture, which lavishes praise on products that have no long-term use or are bought to be thrown away in a moment. The emerging trend of reducing consumption fits perfectly with the ethics of environmental virtues.

4   The virtues of endurance are "the habit of facing danger and difficulties by handling our negative, protective drives and emotions in such a way that we can sustain our chosen sense of place and degree of constructive ecosocial engagement" (van Wensveen 2005, 177). Life is full of challenges, and we must have the strength of character to face them. From these virtues comes the

strength to persist in the intention to protect the environment despite the difficulties that may befall us. In this approach, van Wensveen is reminiscent of Albert Schweitzer, who saw overcoming difficulties as a source of strength and their very occurrence as the inevitable game of life. Among the key virtues in this group, van Wensveen cites (following Randy Larsen) tenacity, describing it as a habit that keeps us between apathy and obsession (1996). Jennifer Welchman (1999), on the other hand, draws attention to the virtue of loyalty.

Louke van Wensveen interprets the catalog in terms of classical philosophy, juxtaposing the four types of environmental virtues with the cardinal virtues that have constituted human's moral compass for centuries, namely prudence, justice, temperance, and fortitude. Van Wensveen asks whether these virtues, known to mankind for centuries, can guide us in the context of humans' moral obligations to the environment. "Has the time perhaps come to supplement or even replace these rusty hinges with something smoothly revolving[11] – with some cardinals that are explicitly environmental virtues?" (2005, 173).

Our daily pursuit of happiness prevents us from seeing the danger of environmental degradation; however, when we face reflection on the state of destruction of our environment, it is such an overwhelming reality that it drives us into despair. Consequently, emotions prevent a person from facing the problem. Hence, looking for new values is not necessarily the best solution. The strongest justification for a specific catalog of virtues comes from data from brain science. According to van Wensveen, our actions in the sphere of morality result from the specific structure of the brain and the mechanisms in it when realizing cardinal virtues and environmental virtues. Each of the four groups of ecological virtues activates the corresponding part of the brain. The same part of the brain is activated when a specific cardinal virtue is put into practice. Thus, van Wensveen links cardinal virtues with ecological virtues. In this way, morality finds its grounding in human biology, thus giving credence to the intuitions of ancient philosophers through recent scientific discoveries.

The author links classical virtues to the aforementioned four groups of ecological virtues, justifying this connection by the specific way the human brain functions. Thus, she finds the basis for the most important cardinal virtues in human biology. According to van Wensveen, such a naturalistic approach, on the one hand, justifies why the cardinal virtues are those from which the others grow, why they have primacy in curbing the most troublesome human vices, and why they are timeless. On the other hand, this approach provides the basis for explaining humans' reactions to certain moral situations. Thus, prudence is linked to the cerebellum as it is responsible for the ability to recognize a situation and adjust for the best emotional response to it in accordance with previously acquired knowledge, taking into account a moral agent's place in a certain structure (Van Wensveen 2000, 182–183, 186). In the ecological era, such a function is performed by the virtues of position, whose function is to adapt our behavior to the specific structure in which we find ourselves. Of key importance here is sensitivity to the environment and the ability to read its needs. At the same time, functioning in the ecological era in the face of the challenges facing humanity requires the virtues of humility, respect, and gratitude (Van Wensveen 2000, 187). These are what help one discover one's place in the universe and maintain the necessary humility, along the lines of the proper humility proposed by Thomas Hill.

The part of the brain associated with the cardinal virtue of justice responds to social situations in a constructive and compassionate manner and enables the experience of grief. The constructive mechanism for responding to certain situations links justice to the ethic of caring, which itself is linked to benevolence and attentiveness. These virtues are considered by van Wensveen to be the most important building blocks of the virtue of justice. Other important virtues that make up justice are friendship and love, which form a just way of responding in certain relationships (Van Wensveen 2000). Van Wensveen emphasizes that the ventromedial prefrontal cortex is responsible for an individual's constructive responses to complex social situations; thus, it combines the virtues of caring with the cardinal virtue of justice.

The dopaminergic system helps to regulate the cardinal virtue of moderation; it also allows a person to direct his actions and desires so as to achieve the superior goals he is pursuing (Van Wensveen 2000, 184), hence the inevitable association with the virtues of attunement – more specifically, simplicity, frugality, and chastity.

The last group of virtues is driven by a completely different mechanism because the virtues of endurance are closely related to the cardinal virtue of fortitude, which helps one to face a problem and be persistent. The correct response to stress is crucial here. Endurance requires fortitude and strength. The occurrence of this moral disposition, according to van Wensveen, is biologically linked to the hippocampus and amygdala.

> The amygdala receives visual and auditory triggers associated with danger and sets into motion a series of reactions that lead to the release of cortisol, which acts both in the body (e.g., by making your heart race, which also makes you *aware* that you feeling afraid or angry) and on the hippocampus.
>
> (Van Wensveen 2000, 185)

In the virtues of endurance, fear for one's safety is replaced by fear for the environment as a whole, and individual courage gives way to courage in concern for our natural surroundings (Van Wensveen 2000, 188). Loyalty here means devotion to the environmental issue and is essential as a virtue that constitutes fortitude. Another key virtue constituting fortitude is tenacity, which is essential in protecting the environment and which prompts us to persist in our resolve to care for it and to continually improve ourselves to do so to the best of our ability.

According to van Wensveen, each group of ecological virtues corresponds to cardinal virtues. They are the modern application of cardinal virtues and they are an answer to the challenges posed by the ecological crisis. According to van Wensveen, the classical virtues are best suited to the ethical challenges human beings faced in the past, as well as those he faces today. The rationale behind their functionality is their grounding in human biology. It is their determination by biology that makes them applicable even today.

3.2.4 Summary

The language of virtue, according to van Wensveen, is therefore both traditional and modern; it is based on certain categories known for centuries in philosophy, but it also refers to new moral challenges. It is an integral, diverse, dialectical, dynamic, and visionary discourse (without social ethics). Although some of the theses put forward by this Dutch researcher are controversial, contradictory, and underdeveloped, her monograph on 'dirty' virtues has played a key role in showing that the issue of virtues is already present in discussions of environmental ethics. Thanks to van Wensveen's book, many ethicists have undertaken to create their own EVE concepts.

Notes

1 The use of the words 'return' or 'restoration' in the context of virtue language implies a return to virtue language in ethics. Virtue is used to analyze the moral dimension in human–environment relations. De facto, virtue language has been in the debate in an implicit form. Hence, this 'return' or 'restoration' rather means bringing "virtues out of hiding" and at the same time refers to the phenomenon of restoring the discussion of virtues in ethics.

2 Louke van Wensveen is an independent ethicist focusing on religious studies, environmental virtue ethics, corporate social responsibility, environmental ethics and sustainability, and religious studies. Born and raised in the Netherlands, she is a graduate of Harvard University (1983) and Princeton University Theological Seminary (PhD in 1987). She worked as a professor at Loyola Marymount University until 2002 and returned to the Netherlands in 2006 to work on introducing sustainable development principles at the Knowledge Centre for Religion and Development. She has also worked in local government (Brummen). For more information on her biography and academic achievements, visit: http://www.dirtyvirtues.nl.

3 Although articles have already been published that address the issue of environmental virtues, e.g., G. Frasz, *Environmental Virtue Ethics: Toward a New Direction for Environmental Ethics*, "Environmental Ethics" 1993, no 3, pp. 259–274; T. Hill, *Ideals of Human Excellence and Preserving Natural Environment*, "Environmental Ethics" 1983, vol. 5, no. 3, p. 211–224, van Wensveen's publication is the first such comprehensive work on the subject.

4 She draws her main inspiration from Murray Bookchin, who is known as the founder of social ecology, which looks for the causes of the ecological crisis in social relations, mainly in human beings' domination over the natural world. Van Wensveen advocates rejecting domination over nature and implementing a harmonious relationship between human beings and the natural world.

5 Berry (2003) developed the so-called concept of the new Earth story, within the framework of which he tried to show humans' moral obligations to the natural environment. He is one of the most recognized representatives of so-called ecotheology, i.e., the current of research that considers ecological issues from the perspective of religion (mainly Christian religions).

6 Feminists believe that the source of environmental degradation is not so much anthropocentrism as androcentrism, which introduced and constituted the privileged position of men, who took neither women nor the natural environment into account in their choices. Androcentrism refers to the conscious or unconscious practice of

emphasizing male points of view and interests and exalting male views and interests over female ones. It is a frequent object of criticism in various feminist theories. In ecological philosophies, it mainly appears when discussing deep ecology and wildlife conservation (Cf. Callicott, Frodeman 2008, 421).

7 Van Wensveen stresses that her exploration of the Aristotelian-Thomistic tradition has helped her understand virtues in the modern world. She interprets virtues and vices through the lens of the four cardinal virtues and the seven deadly sins.

8 The exclusion of the social relations of the moral agent from EVE has been criticized (cf. Rolston III 2005; Hursthouse 2007; Dzwonkowska 2016).

9 The catalog includes virtues and vices excerpted from writings on ecological sociology, deep ecology, bioregionalism, creation theology, animal rights discussions, and ecofeminism (cf. Treanor 2014, p. 39).

10 There is no information about when the author abandoned this line of research, and thus what time range the analyzed literature covers. Nevertheless, in the chapter *Cardinal Environmental Virtues: A Neurobiological Perspective*, van Wensveen writes that she has not engaged in this type of research for more than a decade. Thus, one can conclude that her research ended before the mid-1990s (cf. Wensveen 2005, 176).

11 A play on words, a reference to the original meaning of the word *cardo* (a root of cardinal), meaning "something around which everything revolves, hinges."

References

Anscombe G. E. M., *Modern Moral Philosophy*, "Philosophy" 1958, vol. 33, no. 124, p. 1–19.

Berry T., *The New Story for the Earth: De-mystifying Earth Jurisprudence*, in: *Theilard in the 21ˢᵗ Century: The emerging spirit of Earth*, A. Fabel, D. John (ed.), Maryknoll 2003, 77–88.

Cafaro P., Less Is More, "Philosophy Today" 1998, vol. 42, no. 1, p. 26–39.

Callicott J.B., Frodeman R., *Androcentrism*, in: *Encyclopedia of Environmental Ethics and Philosophy*, J.B. Callicott, R. Frodeman (ed.), Farmington Hills 2008, p. 421.

Dzwonkowska D., *Etyka cnót wobec kryzysu ekologicznego*, "Przegląd Filozoficzny" 2016, vol. 25, no. 2, p. 57–67.

Frasz G., *What Is Environmental Virtue Ethics That We Should Be Mindful of It?*, "Philosophy in Contemporary World" 2001, vol. 15, no. 3, p. 5–14.

Gambrel J. C., Cafaro P., *The Virtue of Simplicity*, "Journal of Agricultural and Environmental Ethics" 2010, vol. 23, no. 1–2, p. 85–108.

Gerber L., *Environmental Virtues and Vices*, doctoral dissertation, New Mexico University 1999.

Gromska D., *Wstęp tłumacza*, in: *Etyka nikomachejska*, Aristotle (ed.), transl. D. Gromska, Warsaw 2007.

Hill T., *Ideals of Human Excellence and Preserving Natural Environment*, "Environmental Ethics" 1983, vol. 5, no. 3, p. 211–224.

Hursthouse R., *Environmental Virtue Ethics*, in: *Working Virtue. Virtue Ethics and Contemporary Moral Problems*, R.L. Walker, P.J. Ivanhoe (ed.), Oxford 2007, p. 155–171.

Jaeger W., *Paideia*, t. 1, transl. M. Plezia, Warsaw 1962.

Kakia, n.d. http://www.theoi.com/Daimon/Kakia.html (Access: February 12, 2022).

Karłowicz J., Kryński A., Niedźwiecki W. (eds.), *Słownik języka polskiego*, vol. 7, Warsaw 1953.

Korpanty J. (ed.), *Słownik łacińsko–polski*, vol.1, Warsaw 2003.

Larsen R., *Environmental virtue ethics: nature as polis*, MA thesis, Colorado State University 1996.

Liddell H. G., R. Scott, H. S. Jones, *A Greek–English Lexicon*, Oxford 1940, http://www.perseus.tufts.edu (Access: January 08, 2022).

Lu M. T., *Hexis within Aristotelian Virtue Ethics*, "Proceedings of the ACPA" 2015, vol. 88, p. 197–206.

MacIntyre A., *After Virtue*, London 1985.

Macy J., *Awakening to the Ecological Self*, in: *Healing the Wounds: The Promise of Ecofeminism*, J. Plant (ed.), Philadelphia 1989.

Nash J. A., *Loving Nature: Ecological Integrity and Christian Responsibility*, Nashville 1991.

Nash J. A., *The Old Order Changeth: The Ecological Challenge to Christian Life and Thought*, "Virginia Seminary Journal" 1997, December.

Newton L., *Ethics and Sustainability: Sustainable Development and the Moral Life*, New Jersey 2003.

Online etymology dictionary, *Arete*, https://www.etymonline.com/word/arete#etymon line_v_16978 (Access: January 15, 2022).

Ossowska M., *Wzór demokraty*, Lublin 1992.

Parry R., *Ancient Ethical Theory*, in: *The Stanford Encyclopedia of Philosophy*, E. N. Zalta (ed.) (Fall 2014 edition), 2014, https://plato.stanford.edu/archives/ fall2014/ entries/ethics–ancient (Access: February 21, 2022).

Plato, *Meno*, transl. B. Jowett, n.d., http://polazzo.com/Plato%20-%20Meno.pdf (Access: February 8, 2022).

M. Plezia (ed.), *Słownik łacińsko-polski*, v. 5, Warsaw 1999.

Rodman J., *The other side of ecology in ancient Greece: Comments on Hughes*, "Inquire: an Interdisciplinary Journal of Philosophy" 1976, vol. 19, p. 108–112.

Rolston III H., *Environmental Ethics: Values in and Duties to the Natural World*, Philadelphia 1988.

Rolston III H., *Environmental Virtue Ethics: Half the Truth but Dangerous as a Whole*, in: *Environmental Virtue Ethics*, R. Sandler, P. Cafaro (ed.), Oxford 2005, p. 61–78.

Shaw B., *A Virtue Ethics Approach to Aldo Leopold's Land Ethics*, "Environmental Ethics" 1997, vol. 19, no. 1, p. 53–67.

Szutta N., *Empiryczna adekwatność Arystotelesowskiej mądrości praktycznej*, in: *W poszukiwaniu moralnego charakteru*, N. Szutta, A. Szutta (ed.), Lublin 2015, p. 171–208.

Taliaferro C., *Virtues and Vices in Religious Environmental Ethics*, in: *Environmental Virtue Ethics*, R. Sandler, P. Cafaro (ed.), New York 2005, p. 159–172.

Tatarkiewicz W., *Historia filozofii*, vol. 1, Warsaw 2014.

Treanor B., *Emplotting Virtue: A Narrative Approach to Environmental Virtue Ethics*, Albany 2014.

Van Wensveen L., *Dirty Virtues: The Emergence of Ecological Virtue Ethics*, Amherst 2000.

Van Wensveen L., *The Emergence of Ecological Virtue Language*, in: *Environmental Virtue Ethics*, *Environmental Virtue Ethics*, R. Sandler, P. Cafaro (ed.), Oxford 2005, p. 15–30.

Węclewski Z., *Słownik grecko-polski*, Warsaw 1884.

Welchman J., *The Virtues of Stewardship*, "Environmental Ethics" 1999, vol. 28, no. 3, p. 411–423.

White L., *The historical roots of our ecological crisis*, "Science" 1967, vol. 155, no. 3767, p. 1203–1207.

Woelke F. A., *Słownik łacińsko-grecko-polski*, Warsaw 1831.

Part II

Contemporary concepts of environmental virtue ethics

Environmental virtue ethics has developed three important concepts, which I will introduce in this part in chronological order. The first is the concept of Henry David Thoreau (1854, Chapter 4), which precedes environmental ethics chronologically, since it was developed as early as the 19th century. The second important concept is EVE by Ronald Sandler (2007, Chapter 5), who owed a huge intellectual debt to contemporary virtue ethicists, basing his analysis on their considerations. The third concept of EVE is narrative ethics as outlined by Brian Treanor (2014, Chapter 6), who drew on a range of philosophical (mainly Martha Nussbaum, Alasdair MacIntyre, and Ronald Sandler). Each chapter describes the ethical concept of the above-mentioned authors (Subsection 1) and the understanding of virtues in this concept of EVE (Subsection 2).

References

Sandler R., *Character and Environment. A Virtue-Oriented Approach to Environmental Ethics*, New York 2007.
Thoreau H. D., *Walden; or, Life in the Woods*, Boston 1854.
Treanor B., *Emplotting Virtue: A Narrative Approach to Environmental Virtue Ethics*, Albany 2014.

4 Classic environmental virtue ethics

The author of the classic concept of EVE is Henry David Thoreau (see: Myerson 1995), who was born on July 12, 1817 in Concord[1] and died on May 6, 1862 in the same town. The residents of this town consider him "their own"; he was attached to his home, traveled little, and led an ascetic life. He was a representative of New England Transcendentalism, which was the literary-philosophical current of his time (Cieplińska 2011, 14; Goodman, Zalta 2017), so his writings should be analyzed from the perspective of philosophy. The ethical, mainly aretological, perspective of Thoreau's thought came from many environmental philosophers. One of the most in-depth interpretations of Thoreau's philosophy comes from Philip Cafaro,[2] and it should be emphasized that interpreting Thoreau's thought is not an easy task because this researcher is more lyrical than systematic (Mooney 2015, 97).

4.1 The foundation of the classic environmental virtue ethics

According to Thoreau, the concept of environmental virtue ethics is derived from New England Transcendentalism, which is considered the predecessor of environmental philosophy and environmental ethics. I refer to Thoreau's concept as "classic" because of its importance in the environmental movement. The term "classic" itself in philosophy has two basic meanings: positive and negative.

> Classic in the negative sense is synonymous with antiquated and outdated. In the second sense, 'classic' denotes something that is outstanding, exemplary, timeless, which, despite changes, is still exemplary, a point of reference, something that is valued despite the passage of time.
> (Morawiec 2004, 23–24)

I refer to Thoreau's concept as "classic" in the sense of a model and inspiration for later environmental ethicists and the environmental movement. This philosopher from Concord is perceived as one of the fathers of environmental reflection in philosophy and literature. His works have inspired generations of environmental thinkers and activists in the United States, from John Muir,

DOI: 10.4324/9781003433156-7

Aldo Leopold, and Rachel Carson to contemporary environmentalists. Hence, from the perspective of environmental philosophy, he is an author fully deserving of the term "classic."

4.1.1 Thoreau as a representative of American Transcendentalism

Although Thoreau is regarded one of the forerunners (Sandler 2006, 135) of the pro-environmentalist movement in the US, he is considered a writer rather than a philosopher,[3] or sometimes a philosophical poet (cf. Mooney 2015, 104) or naturalist; above all, however, he is a representative of New England Transcendentalism,[4] which was born of the romantic inspirations that influenced the Puritan morality of New England. In retrospect, it is clear how great a role this current played in American culture, even though it did not last long (1830–1860): "only for a while there (New England) reigned the intellectual, not the merchant," as Cieplińska emphasizes (2011, 5).

The birth of Transcendentalism in the USA would not have been possible if the intellectual ground had not been prepared for it by Unitarianism. It was this current that forced a rethinking of the narrow dogmas of Calvinism in the light of Rousseau's philosophy, and it was due to Unitarianism that the theory of predestination and eternal damnation was rejected, being replaced by the doctrine of man's perfect nature and God's mercy. This intellectual climate thus opened American thinkers to the Romantic thought of Europe, giving rise to Transcendentalism – a literary and philosophical current, also known as New England Transcendentalism. Transcendentalism was more than a mere adoption of Romantic thought and was formed under the influence of five currents: "Neoplatonism, German idealism and mysticism, Eastern mystical philosophy[5], French utopian doctrines and the Scottish Puritan and Quaker traditions" (2011, 5–6). Neoplatonism inspired the Transcendentalists with its belief in the superiority of spirit over matter; moreover, it instilled a belief in absolute Good, Truth, and Beauty. As Cieplińska writes (2011), German idealism was inspired, among others, by British philosophers Samuel Taylor Coleridge, William Wordsworth, and Thomas Carlyle, all of whom trusted intuition and instinct in place of reason and experience and treated them as tools of cognition; moreover, these philosophers made conscience the source of truth and confidence in oneself.

French utopian doctrines infected transcendentalists with a love for nature that was contrary to the mainstream of the emerging industrial age in the time of Thoreau, which tended to direct affections toward industrialization. It was this love of nature that inspired not only generations of environmentalists, but also figures such as Mahatma Gandhi and Martin Luther King.[6] Moreover, Transcendentalism briefly became the conscience of a nation that was being flooded by a wave of prosperity. Its representatives asked about the quality of life and preached the irreducibility of qualitative criteria to quantitative ones. In a word, the Transcendentalists were a group of idealists who wanted not to improve the material status of the individual but to develop his spirit. With its appeal to the Puritan tradition, the movement took on a moralistic and preachy character, as is very evident in the writings of Thoreau. Like his philosophical teacher Ralph Waldo Emerson (1803–1882),[7] Thoreau was also a harsh critic of American society; or rather, as mentioned earlier, he was the quiet voice of its conscience.

Walden; or, Life in the Woods (1854), along with *Journals* (1837–1861), are considered Thoreau's crowning works. *Walden* is an attempt to show his ideas and express his genuine conviction that the views he professed were possible to put into practice. It is the story of the time (July 4, 1845, to September 6, 1847) that Thoreau spent at Walden Lake on Emerson's land, living in nature, away from the comforts of civilization. The book is a record of his life in harmony with the natural world; it includes a description of building a home[8] and surviving in the simplest and most modest conditions possible. It is a story of a personal journey. Thoreau himself begins his account by saying that he weaves the narrative in the first person. Contrary to generally accepted literary convention, he does so because, as he states, he knows no one else as well as himself. This fact is recognized as a sign of creative authenticity. What he writes in the book flows from experience; it is based not on the accounts of others but on direct knowledge, therefore it is not a reconstructive work but a record of authentic experience.

This type of escape from civilization was not unusual for American transcendentalists; "after all, others, wishing to free themselves from the tyranny of society and escape from industrialization, chose the Brook Farm or Fruitland experiment, for in their view the way to a fuller life was through cooperation" (Cieplińska 2011, 17). Both these initiatives were an attempt to put the idea of Transcendentalism into practice by creating a community that shared the same ideas and followed them in everyday life (cf. Gordon n.d.; Brook Farm 1975). Besides, Transcendentalism itself was a lifestyle in which attention was paid to consistent adherence to the preached principles.

These farms were an expression of the desire to live modestly according to the cycle of nature while adhering to a vegetarian diet, as was fashionable among the Transcendentalists. Thoreau[9] himself did not present a consistent attitude here, for his passion was fishing. He also enjoyed bird hunting, which – despite humanitarian objections – he considered an exceptionally noble and worthwhile sport. In modern times, vegetarianism assumes a consistent stance on these issues and shuns any activity that increases animal suffering, including fishing and hunting. Over time, Thoreau too discovered that fishing and hunting do not go hand in hand with vegetarianism and respect for living beings. As he writes: "The practical objection to animal food in my case was its uncleanness" (ibid, 302). As he goes on to add, repulsion to meat is instinctive in him.

It is difficult to evaluate to what extent Thoreau's isolated form of experimentation was due to his difficulties in social relations. Furtak (2017) even points out that Thoreau's entire intellectual career developed in a polemical relationship with the town where he was born and spent almost his entire life. After his death, the memory of him as a recluse – rather reluctant to meet the residents of the town where he came from – was preserved (Hillway 1945, 328). Although Thoreau writes about the pleasure he took from visitors in his book *Walden; or, Life in the Woods*, he also emphasizes his desire to preserve the space around him. He used to say that one's own thoughts are not always heard in a discussion because great thoughts need space. Perhaps this reluctance to share the farm with others was dictated by the need to focus on great ideas, or perhaps it was a simple expression of selfishness, as some of his biographers suggest (Hillway 1945). Undoubtedly, living in solitude provided Thoreau with space to think, to put his principles into practice, and to test them.

He was undoubtedly an expressive figure, but he could not find himself in his own time – his writings are full of criticism of the society in which he lived. Perhaps this is why he rejected the idea of establishing a community with other Transcendentalists, instead choosing solitude at Lake Walden. Regardless of the circumstances, his description of his experiences during his two years at the lake has become one of the most important works of American literature; researchers emphasize the author's erudition, excellent style, and richness of rhetorical figures. The book is a journey into the natural world, but the author also sees the natural environment as a space for cultivating virtues. This is a biographical item with a strong message (in keeping with the moralistic nature of the American Transcendentalist trend).

In a sense, Thoreau epitomizes in American culture the environmentally virtuous character of the morally outstanding individual who sacrifices everything for a life in harmony with nature. Most environmental virtue ethicists cite Thoreau specifically as an outstanding figure – a so-called "environmental hero." Hence, it is fully justified to take an interest in his views and discover the philosophical layer of his writings. On the theoretical level, an interesting and hitherto unexplored thread is the deontological dimension of Thoreau's thought. His fascination with the concept of dharma provides an interesting starting point for future research on the realization of the duty that this concept imposes. Nevertheless, on an existential level, Thoreau's attitude and narrative of his experiences exemplify the path that leads to moral improvement and becoming the best possible version of oneself through the acquisition of virtue. Therefore, Thoreau could serve as an exemplary moral character (cf. Andersen 2010).

4.1.2 *Thoreau's inspiration from Aristotelian virtue ethics*

Cafaro relates Thoreau's concept to Aristotelian and Romantic thought. He argues that in his writings Thoreau

> notes the danger of knowledge killing off love, science killing poetry, thought killing feeling. Yet *Walden* presents an alternative to romantic despair over the inevitable loss of our initial innocent connection to nature; indeed, it makes a mature, knowledgeable relationship to nature central to living a good life.
>
> (2004, 17)

Walden presents a romantic myth of oneness with nature and an attempt to show the beauty of living in harmony with it. Above all, however, it is a powerful lesson of humility toward nature, for admiration of its beauty is intertwined in *Walden* with the hardships of everyday existence without the conveniences of civilization – an overly humble existence dedicated to observing nature and honing character. Thoreau rejects enjoying the advances of civilization. He writes about abandoning the life of desperation that many people lead, which is an existence without (Latin *de*) hope (Latin *spare*). He encourages living not in desperation but in hope, which leads to recognizing the good in everyday experiences and facing them with courage. For Thoreau, life should be of good quality, reflective and engaged, abounding in experiences that help develop our physical and mental capacities (cf. Sandler 2006, 135). Living in the woods is just that: it is a life lived to the fullest, requiring courage but above all a willingness to experience reality as it is. Thoreau, analogous to the views of the later thinker Schweitzer, says "yes" to life and – like the Alsatian Nobel laureate – accepts it as it is, with its hardships and challenges. He chooses value in place of conformist surrender to quiet desperation. As Cafaro points out, *Walden; or, life in the woods* is a call for thoughtful, reasonable action. Thoreau's actions are guided by reason and forethought, starting with "thoughtfully" building a house, thoughtfully choosing a lifestyle, reading a book, or building a fireplace (Cafaro 2004, 18). All actions are the fruit of thinking them through and looking at them from the perspective of reason, which is, as it were, the realization in life of Aristotle's understanding of virtue.

Thoreau's quest for excellence and human development also brings him closer to the classical understanding of virtue. It is this drive to seek the true experience of life that caused Thoreau to take up the challenge of living in the woods, rejecting the comforts offered by the city. In this he modeled himself on Emerson, who abandoned his well-paid job as a clergyman to seek independent thinking and improve his character, to seek true knowledge rather than be a passive recipient of book knowledge. Emerson's experience was an inspiration for Thoreau, who goes one step further: he gives up the achievements of civilization to polish his character in nature and discover the essence of true life. As Cafaro writes, "Henry Thoreau went to Walden Pond to become a better person, defining this broadly to include increased knowledge, an enriched experience, character development, creative achievement, and greater personal integrity" (2004, 45). He wanted to cultivate virtues, but in a broader, more primordial sense along the lines of the original meaning of the Greek *arete*, a word that was initially equated with strength or fitness, and only later understood as virtue – moral and intellectual. In ancient times, the realization of virtue meant the pursuit of perfection. Being virtuous was synonymous with realizing the fullness of humanity, according to the etymology of the Latin word *virtus*, which comes from the word *vir* (man). Being virtuous means precisely being human in the best sense of the word. This understanding is also indicated by the Greek *arete*, derived from the word *aristos* (best). Nurturing virtue means being the best and as perfect as possible in the full sense of the word. Such a grasp of the concept of virtue may not be refined theoretically, but it has an important function in the practice of life. It signifies a person's striving for moral perfection and shows an example of such action.

The realization of virtue is not so much an end in itself but means following a path to achieve human goals. Virtue is a continuous path that never ends. In Thoreau's view, virtue is not merely a matter of conscience or character but a measure of actual achievement in the world (2004, 48). Virtue is not a dead record that can be contained in a few lines of an ethics textbook or in a poem but requires constant realization in life, and failure to respond to the call of virtue becomes a vice.

Thoreau indicates that we should recover not only the concept of virtue but also the ethical space for nurturing it. This can mainly be achieved by giving economic life its proper place, by setting goals for life, and by persistently trying to achieve these goals (2004, 51). Importantly, Thoreau's views are not exactly a universal ethic that proposes the same qualities and values for everyone. He stresses that every person is different, so we should learn to be independent in choosing our own path in life, following the example of Emerson, who urges not to blindly follow what is written in books, but to think, analyze, and gain wisdom on our own. Thoreau likewise emphasizes the value of individuality and the role of self-development. He admits that we have not only social responsibilities but also the duty to take care of our own development and self-improvement. It should not be overlooked here that Thoreau was accused of neglecting social duties (2004).

Full personal development is constituted by various goods: "health, freedom, pleasure, relationships (interpersonal and natural), experiences (social and natural), knowledge (of oneself and nature) and achievements (personal, professional, social and natural environment)" (Sandler 2006, 135). Thoreau shows that such holistic self-development can have a more beneficial effect on society than philanthropic efforts. It worked in his life, for as Cafaro points out, his contemplation of nature – reflecting on his own experiences and writing the book *Walden; or, Life in the Woods* – gave more to humanity than he could have done in the course of his life, such as visiting the sick or giving alms (Cafaro 2004, 52). Such individual change is the way to transform entire societies. According to Thoreau, the essence of a good life is not philanthropic actions but the pursuit of one's own excellence. He complained that ethics and religion overestimate the value of philanthropy.

4.1.3 Thoreau's inspiration from Far Eastern philosophy

Far Eastern philosophy filled Transcendentalism with mysticism, which, according to Cieplińska, filled the gaps in the system (2011, 6). However, it should be emphasized that Far Eastern thought in Thoreau's case did not so much fill the gaps in the system but provided a background for consideration and influenced Thoreau's thinking about the problems he analyzed. In a sense, the characters from *The Bhagavad Gita* and the content of that epic gave

Thoreau a sense of mission and fulfillment of an individual *dharma*[10] in the forest at Lake Walden. Thoreau himself wrote:

> in the morning I bathe my intellect in the stupendous and cosmogonal philosophy of the *Bhagavad Gita*, since whose composition years of the gods have elapsed, and in comparison, with which our modern world and its literature seem puny and trivial.
>
> (Thoreau 2004, 416)

Thoreau was particularly inspired by the *Bhagavad Gita*, a copy of which he received from Emerson. Cafaro points out that *Walden* has 18 chapters, like the *Bhagavad Gita*, and a similar structure. It begins by showing images of despair and uncertainty, and then the fullness of triumph and affirmation of life in its various guises – in the final pages (2004, 71). The land at Lake Walden is Thoreau's Kurukshetra,[11] where – like the warring Pandavas and Kauravas – the forces of good and evil, virtue and vice clash. Cafaro points to an interesting answer that Thoreau finds in the great epos of Hinduism on the question of duty (*dharma*). Arjuna, one of the Pandava brothers, pondered whether he should fulfill his duty (*dharma*) and take part in a fratricidal battle against the Kauravas. On the one hand, a combination of circumstances made it impossible for him to avoid the war, for it was his duty to take part in it. On the other hand, however, it was a fratricidal battle: if he had entered it, he would have had to kill his relatives and teachers. Faced with these doubts, Arjuna asked his master Krishna for advice. Krishna, in turn, presenting an answer based on Hindu philosophy, explained to Arjuna that fighting evil was his *dharma* (duty). By fighting evil, he upholds the sacred *dharma*, thereby helping to save the world. This message from the *Bhagavad Gita* sets Thoreau on the path to recognizing the importance of fulfilling the *dharma* of one's own life, understood as fulfilling the duty one has been assigned – striving to realize one's vocation. Thoreau points out that different people may have different duties or vocations to fulfill in life. What's more, a given period in one's life may involve different duties, depending on one's age, position, and situation.

As a perfect example of devotion to duty and conscientious performance of obligations, in *Walden* Thoreau gives the example of an artist from Kouroo who decided to make the most perfect scepter possible. He went to the forest in search of the best tree. During this time, his friends gradually deserted him as they grew older, while he himself did not age a bit.

> His singleness of purpose and resolution, and his elevated piety, endowed him, without his knowledge, with perennial youth. As he made no compromise with Time, Time kept out of his way, and only sighed at a distance because he could not overcome him.
>
> (Thoreau 2004, 456)

The search for the trunk for the scepter took so long that the city of Kouroo had fallen into ruin; before he gave the trunk he found the desired shape, the Kandahar dynasty was extinct; before he smoothed and polished the scepter, the *kalpa*[12] was over, and before he carved the ornament and head of the scepter, many Brahma days and nights had passed. The artist was completely absorbed in his work,

> when the finishing stroke was put to his work, it suddenly expanded before the eyes of the astonished artist into the fairest of all the creations of Brahma. He had made a new system in making a staff, a world with fun and fair proportions in which, though the old cities and dynasties had passed away, fairer and more glorious ones had taken their places. And now he saw by the heap of shavings still fresh at his feet that, for him and his work, the former lapse of time had been an illusion, and that no more time had elapsed than is required for a single scintillation from the brain of Brahma to fall on and inflame the tinder of a mortal brain.
>
> (Thoreau 2004, 457)

Many researchers are inclined to attribute authorship of this story to Thoreau, who undoubtedly identified with the artist from Kouroo, and his work was *Walden; or life in the woods*. Thoreau stressed that, like Arjuna, he had to do his duty, even when, as in the *Bhagavad Gita*, it seemed to conflict with generally accepted social norms. Krishna's lesson is a teaching on *dharma* and dedication to carrying it out. So, Thoreau worked on his moral character, and during his sojourn at Lake Walden he saw his path and the resulting message for humanity. The figure of Arjuna and his experience in the field of Kurukshetra were an inspiration to Thoreau. The lesson from Krishna's teachings, emphasizing the importance of duty and the importance of fulfilling one's *dharma*, recurs repeatedly in Thoreau's writings. Besides, the reference to duty itself is peppered with martial metaphors, so to speak, lifted from descriptions of the *Bhagavad Gita*. Emerson emphasizes that there was something of the artistry of war in Thoreau's nature, in his way of striving for the realization of virtue. At the same time, he demonstrated with his life a commitment to the duty of self-improvement.

4.1.4 Summary

The theoretical foundation of Thoreau's thought is not impressive, but his book about his sojourn at Lake Walden is a story of his own experience. The book could be an example of a description of an individual's way of pursuing virtue, along the lines of Aristotelian theory (as Cafaro suggests), or it could show a private story of pursuing one's own *dharma*.

Although Thoreau's philosophy is not theoretically elaborate, his role in the discussions taking place in the field of environmental virtue ethics is so important that it would be a mistake to overlook his thought. This philosopher from

Concord has become an icon of the environmental movement and environmental pop culture.[13] Above all, he is an unsurpassed model of paradigmatic moral character. Hence, in his conception, more important than the theoretical underpinning is the practice and ethos of the ascetic nature lover. I will present this in the next section, which is devoted to the virtues most important to this American thinker.

4.2 Virtues in classic environmental virtue ethics

Thoreau's concept is less theoretical than the other EVE concepts, being rather a lyrical description of his own experiences – a proposal of a kind of a parsimonious and radical ascetic ethos. Although the rejection of civilization seems from the perspective of modern man to be an extreme action, for Thoreau it was a manifestation of fidelity to his own views, as taught to him by Ralph Waldo Emerson. It was Emerson who, in August 1837, at the inauguration ceremony of the academic year at Harvard University, gave a speech that became an inspiration to many young people (cf. Cafaro 2004, 1). In that speech, Emerson criticized the unreflective following of bookish knowledge as leading young people astray because it does not allow them to think independently and analyze reality. Independent thinking is essential for discovering one's own individual vocation and a lifestyle that serves the realization of the highest ideals.

Thoreau's vocation was pursuit of moral perfection during his stay in the woods at Walden Lake. Emerson expressed his opinion as a person who had turned down a well-paid job five years earlier to develop an independent style of thinking. In this way, he gave credence to his spoken words with his attitude and encouraged young people to seek their own path in life and to develop virtues in the classical sense of the word, that is, by improving their own character. He encouraged heroic effort on one's own path in life, effort in the pursuit of moral perfection, perfection of character. This exhortation was a coherent message from a person who himself made such an effort to act in accordance with his own convictions. Thoreau, moreover, believed that philosophy is not about sophisticated thinking or even the creation of philosophical systems. A philosopher, according to him, is one who loves wisdom enough to live a simple, independent, generous, and trusting life in accordance with the precepts of his beliefs (Cafaro 2004, 36). Therefore, *Walden; or, Life in the Woods* is not a universal set of principles but a description of an individual implementation of the principles of philosophy in life, inner inquiry in the style of Socratic self-knowledge, and the pursuit of moral perfection. According to Cafaro, Thoreau's romantic view of self-improvement adds an individualistic aspect and authenticity to the ancient concept of virtue.

Thoreau, like many other Transcendentalists, did not want his philosophy to be just a useless theory. He wanted it to be an authentic testimony to his life, as he proved when he moved to the woods at Walden Lake, and when he refused to pay taxes[14] in line with his views on agreeing to disobey the state when it acts improperly. Thoreau was consistent from beginning to end in his pursuit of moral excellence, practicing it surrounded by nature. As has been shown more than once in this work, he saw in nature a reflection of the ideal to which he aspired. To his adoration of the natural world he tied his extensive knowledge of Greek and Roman mythology, poetry, philosophy, religion, Far Eastern spirituality, and culture. It was no different when he cited *Gulistan or Rose Garden* (1865) by the Persian poet Saadi of Shiraz (1213–1295), who says in this work that the cypress is a noble tree, even though it does not bear fruit. This is because the nobility of this tree comes from the fact that it is always fresh and lofty, unaffected by the seasons. This is precisely the morality of man, who should always preserve what is noblest in him, regardless of changing external conditions. The concept of flourishing in the moral sense "suggests beauty of character and higher achievements than mere physical endurance or physical growth" (Cafaro 2004, 21).

Thoreau was accompanied by the Romantic concept of *Bildung* (Cafaro 2004, 23), an idea spread by Kant, for whom the Enlightenment is seen as man's emergence from immaturity (Bałżewska 2012, 112). *Bildung* was a central concept also for many transcendentalists, including Emerson (Cafaro 2004, 23). For Thoreau, the way to self-improvement was to participate in everyday trivial activities. At the same time, imbued with moral ideals from philosophy, religion, and spirituality, it was in these simple activities that he saw a deeper dimension and opportunity for moral improvement. Tilling the land, getting up at dawn, bathing in the lake, or watching the sunset are the backdrop for the deep moral reflections and thoughts we find in the pages of the book under review. They are an expression of the constant search for quality of life rather than its material dimension. It is moral perfection that is the basis of the doctrine of simplicity (Stoller 1956, 458), through which Thoreau seeks to discover the deepest meaning of life, untainted by trivial matters.

It was precisely for this purpose that Thoreau changed his place of residence and settled at Walden Lake. As he writes:

> I went to the woods because I wished to live deliberately, to front only the essential facts of life, and see if I could not learn what it had to teach, and not, when I came to die, discover that I had not lived. I did not wish to live what was not life, living is so dear; nor did I wish to practice resignation, unless it was quite necessary. I wanted to live deep and suck out all the marrow of life, to live so sturdily and Spartan – like as to put to rout all that was not life, to cut a broad swath and shave close, to drive life into a corner, and reduce it to its lowest terms.
>
> (Thoreau 2006, 128–129)

These words can be considered Thoreau's credo; they can be taken as the guiding idea for his experiment, the essence of which is the desire to discover life of the best possible quality, life that is simple on the one hand, and so bluntly true on the other.

Thoreau recognizes that the concept of virtue differs from that of a character trait in that virtue contributes to the flourishing and development of a person. In Thoreau's thought, it is freedom, moderation, and respect for nature that seem to be the greatest virtues. Nevertheless, according to Cafaro, there are many more virtues in Thoreau, and they can be divided into six groups, depending on what sphere of life they concern. These are:

1 "Personal virtues: help us act effectively and own our actions. (…)
2 Social virtues: foster good relations with others and avoid immorality. (…)
3 Intellectual virtues: contribute to knowledge of the world around us and successful action within it. (…)
4 Aesthetic virtues: further the creation and appreciation of beauty in art and nature. (…)
5 Physical virtues: facilitate physical activity, health, and well-being. (…)
6 Superlative virtues: promote or mark extraordinary human excellence"

(Cafaro 2004, 57–58)

Cafaro describes Thoreau's list of virtues as artificial. He points out that some of the virtues listed fall into more than one area. Nonetheless, he captures well Thoreau's idea that virtue should be related to progress in specific spheres of life, each of which is referred to by one of these categories. However, the most important cardinal virtues for Thoreau are moderation and simplicity, which lead to freedom and respect for nature. It is the possession of these virtues that is particularly important to Thoreau's individual personal development and forms the ecological ethos in his philosophy.

4.2.1 *Moderation and simplicity*[15]

Two of Thoreau's most important ideals were modesty and simplicity. In turn, he criticized people's hoarding of "treasures which moth and rust will corrupt, and thieves break through and steal" (Thoreau 2006, 8). He himself chose a very ascetic life based on the model of nature. As he wrote: "every morning was a cheerful invitation to make my life of equal simplicity, and I may say innocence, with Nature herself" (Thoreau 2006, 124). Nevertheless, this asceticism was natural to him: it did not bother him, it did not challenge him, except perhaps for the intellectual challenge, within which reason set its framework. Thoreau also wondered what the limits of our needs are; he recognized that it is necessary in the first place to have food, shelter, fuel, and clothing (Thoreau 2006, 18). But what does it mean to provide basic conditions for living? What is the framework of optimal conditions? Thoreau recalls Darwin's tale of the inhabitants of Terra del Fuego. Darwin's traveling companions, despite being warmly dressed and sitting by the fire, were freezing, while the natives, who were naked and away from the fire, were oozing with sweat. Similarly, he cited the example of the citizens of New Netherland, who could walk around naked while Europeans shivered from the cold (Thoreau 2006, 19). Thoreau thus asks: "is it not possible to combine the fortitude of the savage with the intellect of civilized man?" (Thoreau 2006). That is, to strive for an adequate body temperature, with an ascetic minimum, quite in the style of Thoreau, who is constantly guided by the idea of satisfying needs without unnecessary luxuries.

Moreover, Thoreau stressed that clothing should only protect the body from the cold and, for cultural reasons, cover up nudity. He criticized succumbing to fashions and acquiring clothing that we could easily do without when we are "led oftener by the love of novelty and a regard for the opinions of men, in procuring it, than by a true utility" (Thoreau 2006, 31). In his opinion, this detachment of clothing from its original functions does not serve human development. He himself, as he stresses, believes that there is nothing wrong with wearing patched clothing if it still fulfills its functions. As he proves more than once during his stay at Lake Walden, sparing resources and avoiding waste are his top priorities. What's more, having fashionable clothes is one symptom of being overly concerned with secondary things instead of what is important in life. Thoreau writes: "I am sure that there is greater anxiety, commonly, to have fashionable, or at least clean and unpatched clothes, than to have a sound conscience" (Thoreau 2006, 32). He consistently stressed that material things are secondary and that one should strive first and foremost for higher goals, whereas many people focus on unimportant things: "for the most part we allow only outlying and transient circumstances to make our occasions" (Thoreau 2006, 189).

According to Thoreau, clothing has become an expression of blindly following fashion and a way of emphasizing the material status of the person who wears it; meanwhile, a change of clothes is only necessary when we as individuals change, just as happens in nature, where the snake sheds its skin due to growth, and the caterpillar becomes a larva. The change occurs because clothing is the "outmost cuticle and mortal coil" (Thoreau 2006, 35). Clothing should be as simple as possible; in terms of possessions, the number of items should be kept to a minimum.

What else does a person need to function? Thoreau lists a few simple items, such as knives, axes, shovels, or wheelbarrows; and for people engaged in science, a lamp, stationery, and books (Thoreau 2006, 20). In his view, additional conveniences are unnecessary and even harmful. They are a serious obstacle to the spiritual life of humanity. As this ascetic points out:

> With respect to luxuries and comforts, the wisest have ever lived a more simple and meagre life than the poor. The ancient philosophers, Chinese, Hindoo, Persian, and Greek, were a class than which none has been poorer in outward riches, none so rich in inward
>
> (Thoreau 2006, 21)

Thoreau makes it clear that the pursuit of material wealth should not become a man's goal; moreover, the possession of excessive wealth is a yoke that enslaves man. He himself saw "young men, (…) whose misfortune it is to have inherited farms, houses, barns, cattle and farming tools. (…) How many a poor immortal soul have I met well-nigh crushed and smothered under its load" (Thoreau 2006, 33–34). He saw possessions as a yoke to the soul, and the need to deal with them as an obstacle to the soul's quest for liberation. As he believed, the preoccupation with worldly possessions made people preoccupied with imaginary problems instead of what was important: the cultivation of a beautiful soul or – as he writes in other passages – the pursuit of awakening the divinity of the human interior, liberation from self (Thoreau 2006, 11). These postulates of awakening the innate divinity or liberation from self are a reference to Far Eastern philosophy. Perhaps also the portrayal of the charioteer, who is ignorant of eternal matters – too focused on temporal matters – is a subtle allusion to Arjuna, who, in addition to the principles of war, also presents to his disciple the principles of liberation from the shackles of one's own smallness and the realization of innate divinity, as well as the throwing off of the golden or silver shackles of duty with which man has shackled himself in the name of acquiring superfluous luxuries.

Modern construction was also resented by Thoreau. Here, too, he was a proponent of the simplest solutions, meeting only basic human needs. Moreover, he criticized the low availability of housing for people in cities and the high rent paid for the opportunity to live in a house that one does not own. Again, he gives the natural world as an example, where every animal has its own place of existence. For him, the second example was the world of uncivilized people (as he described the Native Americans), whose superiority lay in the fact that they sufficiently meet their simplest needs (cf. Thoreau 2006, 43). Thoreau also criticized access to housing: he pointed out that money from rental housing could buy a village of wigwams. Unfortunately, this money is used to keep Native Americans in poverty for the rest of their lives. He criticized a civilization that is so organized that people cannot afford to buy or build their own homes.

However, being a civilized person means bearing the costs of renting an apartment, or possibly living with debt if you decide to buy your own property or farm. The complicated mechanisms of modern economics mean that the paradox of living a life of luxury yet deprivation creeps into the lives of ordinary people. On the one hand, we have access to greater comforts than uncivilized people, but on the other hand we are deprived of a thousand pleasures that they had. Moreover, we are enmeshed in market mechanisms that make it necessary for us to solve the problem of access to basic goods by means more complicated than these problems. "To get his shoestrings he (the farmer) speculates in herds of cattle" (Thoreau 2006, 48). The development of civilization does not go hand in hand with the enrichment of people.

Nevertheless, the achievements of civilization make life easier for humans. One should not get the impression that Thoreau wants people to return to primitive conditions – to living in caves or wearing skins. "It certainly is better to accept the advantages, though so dearly bought, which the invention and industry of mankind offer" (Thoreau 2006, 58). They are the fruit of experience and the accumulation of knowledge to develop man and improve his well-being. Thoreau defines civilized man as a more experienced and wiser savage who harnessed knowledge and economics to make his life easier. He criticized the way economics and social life were organized. He was averse to the consumerist drive of American society. In a sense, he was an anti-technology luddite,[16] but at the same time he claimed that the value of the gains resulting from development cannot be completely rejected but should be adjusted so that they do not obscure the value of living a qualitatively and morally good life. With his life he confirmed the thesis that obtaining the food necessary to feed an individual requires improbably little effort, and that simple food is enough to keep a person healthy and strong. He fed himself in an exceedingly modest way: meals were usually prepared outside on the fire, even in the rain.[17] The description of his diet (Thoreau 2006, 88–93) is an expression of his philosophy of simplicity and modesty as it shows how he baked the simplest cakes from a mixture of rye and corn flour without the use of yeast, which he considered unnecessary. He ate what he grew himself, even getting a surplus crop to sell. The only seasoning he used was salt in small amounts, but, he claimed, he could do without that as well. The cost of additional food, such as flour, salt, and sugar, amounted to only \$8.74 for eight months, a very modest expenditure despite the lower purchasing power of the dollar and a testament to his clear reduction of needs to the bare minimum.

Thoreau wanted to show that the fact that we work so much is because we earn money for things we don't really need. While at Lake Walden, he proved that from the work of his own hands he was able not only to feed himself but also obtain a surplus crop that he could exchange for other necessary goods. He argued that man spends too much time satisfying unnecessary needs, and it would be enough to work about six weeks a year to get what is necessary for survival. At a graduation ceremony at Harvard University, Thoreau gave a speech in which he encouraged the rejection of the order of working six days a week and one day of rest, which stems from Bible. In his view, this order should be reversed, meaning one day of work and six days of rest (cf. Cieplińska 2011, 87).

The experiment at Lake Walden confirmed his words. He chose to live a very simple life, in terms of not only the food he consumed but also furnishing his home. All he had at home was a bed, a table, a desk, three chairs, a mirror, a pair of tongs, a kettle, a saucepan, a frying pan, a brewing pan, a bowl, two knives and forks, two plates, one cup, one spoon, a kerosene jug, a molasses jug, and a lamp (Thoreau 2006, 93–94). Why so little? Because more is not needed for a person to function well. In addition, the more things we have, the poorer we are. He believed that caring about material things consumes too much time and distances us from the important things. Above all, he praised reducing the number of things we own. He wrote: "cultivate poverty like a garden herb, like sage. Do not trouble yourself much to get new things, whether clothes or friends. Turn the old; return to them. (…) Superfluous wealth can buy superfluities only" (Thoreau 2006, 458–460). One should take care of the inside, one's own moral and intellectual development, not the accumulation of things. We need things only in the minimum amount necessary for survival; we do not need to accumulate them.

Moreover, Thoreau's modest possessions during the Walden Lake experiment clearly bypass the most important element of capitalism – money (Schneider, Myerson 1995, 98). Thoreau's actions seem to have been directed more often to barter than to the use of this means of payment. Money was not necessary for him to live on when he kept his needs to a minimum and exchanged the surplus for what was necessary for survival. He gave the impression of a person who discovers precisely in such restriction what is most important – true freedom.

4.2.2 Freedom

By reducing the number of belongings, one moves toward true, unfettered freedom – a freedom from things, from social obligations, and from the demands of the times – which allows the individual to enjoy what matters, namely genuine life. Things are like trinkets attached to a belt that hinder our movement – like a prison for a person.

> If you are a seer, whenever you meet a man you will see all that he owns, ay, and much that he pretends to disown, behind him, even to his kitchen furniture and all the trumpery which he saves and will not burn, and he will appear to be harnessed to it and making what headway he can.
> (Schneider, Myerson 1995, 95)

The possession of things enslaves. First, by harnessing vital forces to acquire them; second, by the need to care for and maintain them; and third, by the compulsion to provide protection for things out of fear of theft. Thoreau wrote of feeling pity for "compact-looking people" who, ready to go, worry whether they have insured their property (Schneider, Myerson 1995). Possessions enslave a person, they are like a spider's thread that entwines, preventing him from enjoying his freedom. Thoreau condemned the accumulation of things because he recognized that most of them are unnecessary junk that we don't know how to throw away, although we should get rid of them. Things take away our freedom, time, and space. Thoreau kept furnishings to an absolute minimum; he didn't even opt for curtains, recognizing that when he was bothered by excessive sunlight, he would hide somewhere rather than expand his home furnishings by one more thing (Schneider, Myerson 1995, 96).

Having possessions can be likened to a person traveling with a great deal of baggage – bending under the weight of his own "trash." Thoreau says this about those who carry their possessions:

> when I have met an immigrant tottering under a bundle which contained his all – looking like an enormous well which had grown out of the nape of his neck – I have pitied him, not because that was his all, but because

he had all that to carry. If I have got to drag my trap, I will take care that it is a light one and does not nip me in a vital part. But perchance it would be wisest never to put one's paw into it.

(Schneider, Myerson 1995)

Possession of things, according to Thoreau, is a snare as it restrains man and takes away what he holds most precious – freedom and independence; this is one of Thoreau's most important lessons. As for work and rest time, Thoreau resolutely limits work time to that which is used to obtain the things necessary for survival. By replacing six days of work with one working day, one gains time to engage in contemplation of nature and study, which frees the body and mind. The body is freed not only from the excess of gainful work, but also from the responsibilities of taking care of possessions. The mind is freed from caring about possessions, and from thinking about wanting other things.

Thoreau relieves a man of the excess of social relationships he would get caught up in by owning things, for example, to earn money for them, insure them, take care of them (when something needs professional care). Cafaro writes that enslavement to desires is worse than enslavement to people. Unfortunately, we impose this worse form of enslavement on ourselves.

Such slavery is worse than chattel slavery, in the sense that it is wholly self-imposed. We cannot blame it on anyone but ourselves. Escape from such slavery depends on *thinking* our way out of it, for 'what a man thinks of himself, that it is which determines, or rather indicates, his fate'.

(Cafaro 2004, 42)

Anyway, the book *Walden; or, Life in the Woods* begins with a complaint about the enslavement that attachment to land brings to man. Thoreau writes: "I see young men, my townsmen, whose misfortune it is to have inherited farms, houses, barns, cattle, and farming tools; for these are more easily acquired than got rid of" (Cafaro 2004, 8–9). He goes on to ask the question, "who made them serfs of the soil?" (Cafaro 2004). Their misfortune stems from the fact that they have more land than they need to feed themselves, and at the same time they have become part of a certain machine that has made them people with burdens that take away their freedom. They are overwhelmed by the responsibilities that come with owning such large estates, and their task can be likened to an Augean stable, which is a depressing enough metaphor to show how great their enslavement is. Thoreau criticizes such enslavement, calls it the life of fools, and is surprised that in this, as he puts it, relatively free country, people have imposed such a burden on themselves that they focus on imaginary concerns instead of on the essence of life (cf. Cafaro 2004, 9–10).

Perhaps this was a reference to the Far Eastern doctrine of the soul's enslavement in the body, where, following the pattern of Orphic beliefs, the soul transcends the mortal body, and in a sense, in both these concepts – Far Eastern and Orphic – the body is the soul's temporary home, even a prison. The reference to Far Eastern beliefs is also found a few verses further on, where Thoreau speaks of the divinity of the human interior, and then of the coachman from his town, in whom this divinity does not awaken. On the contrary, he is trapped in a quagmire of daily duties, fear, and his own opinion of himself. The reference to the charioteer may not be coincidental; it is well known that Thoreau was fascinated by the *Bhagavad Gita*, in which the metaphor of the charioteer is one of the most important. Thoreau may have deliberately chosen the charioteer from his town to show the fallacy of human life – enslaved by daily chores, detached from thinking about what is important. Above all, however, the key is the detachment from oneself, from what the individual thinks about himself, from his own opinion of others. As already mentioned, what we think creates our fate. What a person thinks about himself can enslave or liberate him. It was the life of the charioteer – or the enslaved life – that Thoreau referred to as a life of quiet despair (Cafaro 2004, 10).

What liberates is the intellect (cf. Cafaro 2004, 42), just as in the *Bhagavad Gita* the intellect is described as the source of both enslavement and liberation. As Thoreau showed, people place the burdens of imaginary cares and problems on themselves. According to Cafaro,

> Thoreau sees true freedom in the play of the intellect itself, whether in personal expression and artistic creativity, in the detailed and precise observation and reasoning of science, or in the planning and building of a house or a new pencil-making process.
>
> (Cafaro 2004)

It is freedom that rises above trivial, mundane matters in the encounter with great thoughts. Freedom is hidden in the great ideas of the intellect, but also in very mundane activities. However, it is, above all, the freedom of the mind unfettered by an excess of duties; it is more than the freedom of the body: it is the true freedom of the human spirit, the practical expression of its greatness and supremacy over the body.

Thoreau also paid attention to the enslavement of the body. He opposed slavery. In his opinion, everyone deserves freedom. However, for him, freedom meant not only the absence of physical compulsion, but also the intellectual openness and ability to nurture one's own thoughts (cf. Cafaro 2004, 30). After all, thoughts, especially great ones, need space to thrive. Thoreau sometimes complained about being too close to his interlocutor. Thoreau wrote:

> one inconvenience I sometimes experienced in so small a house, the difficulty of getting to a sufficient distance from my guest when we began to utter the big thoughts in big words. You want room for your thoughts to get into sailing trim and run a course or two before they make their port.
>
> (Cafaro 2004, 198)

When the conversation descends into trivial topics, the proximity of the interlocutor is justified, but when it concerns important matters, then a certain distance is advisable so that the thoughts can properly resound. Thoreau appreciated solitude. It was solitude that gave him the space to think. A certain distance from people was an expression of his freedom from them, but it was also a life necessity for every individual. Thus, freedom is not only the absence of enslavement to the body, but also intellectual freedom, freedom to think, and space so that great thoughts can be born. For Thoreau, full freedom also means freedom of thought and imagination and is the key to making free choices (Cafaro 2004, 215).

The pursuit of freedom was also important for other Transcendentalists, including Thoreau's teacher, Emerson, who believed that creative thinking was necessary for our intellectual independence. In a speech at the beginning of the academic year, cited earlier, he stressed the importance of independence in thinking – the need to discover what is good for the individual, rather than duplicating general standards or mindlessly repeating book knowledge.

> We must think our own thoughts. Not, of course, that we will fail to take advantage of past discoveries or study the world's literary, scientific, and religious traditions. But we will interrogate the ideas of the past and put them to the test of our own experience.
>
> (Cafaro 2004, 2)

Such is the idea of freedom that can be seen in Thoreau, who did not succumb to the influence of others, and who used broad erudite knowledge as a background for thinking through his own experiences, for seeing certain analogies between the natural world and, for example, mythology, philosophy, or poetry. This was exactly what Emerson was talking about: using one's own knowledge of the ideas of the past to observe reality and create one's own reflections based on individual experience.

It seems that freedom is a value in itself; it is one of the foundations on which Thoreau's ethos is built. Unlike moderation, which is merely a value that leads to a goal, freedom is a value that is a continuous path to self-improvement. It is modesty, restraint of desires or moderation that leads to the attainment of this coveted value of freedom, and from the practical side it makes a person independent. Freedom, according to Cafaro (2004, 62), is a key value in Thoreau. "Freedom is an important goal, but we may dishonor freedom through personal swinishness as well as through unjust actions toward others" (Cafaro 2004, 174). Freedom requires us to be upright and improve ourselves morally; the preservation of freedom is entirely dependent on our conduct and behavior toward others.

According to Cafaro, freedom occurs in Thoreau in two forms: positive and negative. Negative means man's freedom from desires, economic needs, social relations, fear, and desperation. The Walden Lake experiment is an example of the implementation of this very form of freedom. But alongside this negative freedom, which seems to be the first step on the road to true liberation, there is also positive freedom, which is the unfettered ability to develop one's own talents, moral prowess, thoughts, and artistic work; in general, it is the freedom to live one's own life (Cafaro 2004, 62). While negative freedom seems to be a value – an ideal to which one strives – positive freedom is a virtue, an ethical fitness in which a person excels. Positive freedom is the independence to develop what is best in the moral agent, which allows him to discover the fullness of his humanity, to be the best version of himself. "Negative freedom does not appear to be an end in itself for Thoreau. It finds value and completion in positive freedom: the full flourishing and expression of individual personality" (Cafaro 2004). It is the fulfillment of Emerson's exhortations to live one's own life, to discover the individual purpose of one's own life, for this constitutes the culmination of efforts in the pursuit of full freedom.

4.2.3 Respect for nature

Nature is the background for Thoreau's journey into the world of the austere, ascetic life,[18] which frees him from social and economic bonds, making him a free man but also a morally better one. The description of his inner journey is laced with many rich images of the natural world. Although Thoreau played a greater role in the formation of ecological currents than as a moralist, it should be remembered that nature is only the background for the essence of his quest, albeit an extremely important background that is of great importance in the pursuit of the two virtues described previously. The third virtue crucial to the realization of the ecological ethos as seen by Thoreau is, in my opinion, respect for nature.

Thoreau writes the word "Nature" with a capital letter. This shows the great respect Thoreau had for the natural world. He referred to Nature as the old lady who resides in his neighborhood: "An elderly dame, too, dwells in my neighborhood, invisible to most persons, in whose odorous herb garden I love to stroll sometimes" (Thoreau 2004, 194). For him, Nature is indescribably innocent and benevolent, and her gifts – sun, wind and rain, summers, and winters – give unparalleled health, strength, and joy. In a sense, human beings are connected to nature. It can be argued that Thoreau was a forerunner of thinking about human–environment relations in a way that is close to the idea of the network of connections that we find in Leopold's biotic community or the concept of the interconnectedness of man and the environment, of a certain empathy with nature that is inherent in deep ecology. For Thoreau, a human being "is made of leaves and plants," which is a clear sign that people are part of Nature. Human beings are created from the same ingredients as other natural entities.

Nature provides us with the best that it has to offer, and at the same time it is in its resources and wisdom that the cures for all diseases are hidden.

Nature is the source of health, providing a plant-based, universal remedy for all human ailments that is more effective than any other specifics (Thoreau 2004, 195). Moreover, it is also a life-giving force, a source of vitality and longevity. This is evidenced by the fact that nature and the natural world have outlived even the oldest recorded inhabitants of our planet. The universally available goods of nature are a source of health and vitality, like the morning air, which can replace all the medicines distributed by carriages. Thoreau writes of himself as one who worships Hebe, the Greek goddess of youth who restores vitality to gods and men, who on Olympus served nectar and ambrosia to the gods and was the personification of strength, health, and beauty, a reflection of perfect physical form. Wherever she appeared, spring began (Thoreau 2004, 196).

Thoreau was so fascinated by nature that he felt a genuine joy in being in it. He subsisted on the work of his own hands and his own crops, but it gave him undying joy. In his own words, he drew strength from cultivating the land (Thoreau 2004, 218). Nature was his teacher, and he observed it to gain knowledge about the cultivation of plants and the processes of life. His observations were accompanied by admiration for nature and appreciation of the work of his own hands, which, he claimed, has an imperishable moral sense (Thoreau 2004, 221), for the astute observer learns the true life in its wonder and simplicity. The processes of nature reflect those processes to which man is subject. Just as our body is constructed in imitation of the various elements of the natural world,[19] so the natural world resembles man's journey toward perfection. In his description of spring and waking life, Thoreau compares sprouting plants to the goodness and virtues sprouting in man. On the one hand, everything is being pulled apart so that the sprouting plants can break through to the surface; on the other hand, under the influence of the evil done during the day, the seeds of virtue are hiding (cf. Thoreau 2004, 441).

The soil works in ways familiar to a man who is living a moral life. Following John Evelyn, Thoreau argued that the earth has a certain magnetism in it that gives life and attracts salt, strength, and virtue. However, it requires human labor and effort. Today it has become merely a source of income and is treated with haste and carelessness. It has been denied the right to be a hallowed art, as ancient poetry and mythology treated it (Thoreau 2004, 233–235). The ancient world celebrated festivals associated with giving thanks to the earth for crops and abundance. Offerings were made to Ceres and Jupiter, while today the land has become nothing more than a source of hard labor, and its possession is the fruit of stinginess and greed. Cultivation of the land[20] is not what it once was, and the farmer leads a miserable life; moreover, treating the land as a means of acquiring property has led to the deformation of the landscape. Thoreau cites ancient authors for whom land had more value than it does today. Cato was said to have proclaimed that the profits from agriculture were of a particularly pious nature, and the ancient Romans called the earth their mother and held those who cultivated it in high esteem. As you can see, respect for nature and the natural world was not limited to wild forms of life but included the world of agriculture, which stands at the intersection of nature, the natural world, and artifacts. In ancient times, the work of a farmer was held in high esteem. Besides, Thoreau himself had a special love for simplicity and vitality, as can be seen from his enthusiastic remarks about a Canadian lumberjack (Thoreau 2004, 203–221) who visited him. His simple lifestyle was what he himself wanted to show with his life; a gentle disposition and contentment with life was the best recommendation for a simple, modest life.

Nature is the best teacher of the virtues and ideals to which man should aspire. Those who spend a lot of time in nature can derive more knowledge from it than is gained by a scholar who spends his days in a library.

> Fishermen, hunters, woodchoppers, and others, spending their lives in the fields and woods, in a peculiar sense a part of Nature themselves, are often in a more favorable mood for observing her, in the intervals of their pursuits, than philosophers or poets even, who approach her with expectation.
>
> (2004, 297)

Nature teaches mankind valuable lessons – teaches wisdom and virtues.

The sun, for example, teaches mankind the lesson of treating everyone equally, for its rays fairly divide all the meadows and fields on which they fall. It also teaches us dedication to our tasks. Every day it does its work. It does not single out one or depreciate another; it falls equally on the good and the bad.[21] The sun teaches us to respect the tireless performance of our own duties. Out of reverence for nature, Thoreau took with gratitude what others took for granted. For him, the sun's rays were more than a phenomenon experienced daily. Almost like a child, he was in awe of all natural phenomena; his admiration came from the sun, the wind, the shade, the clear water in the pond, the luscious green of the forest, the color of the sky. Everything that came from this noble old lady filled him with awe that was even disproportionate to the mundanity of these phenomena.

Nature reveals its essence to those who spend time in nature, not to those who learn it from scientific works or through travel. "He who is only a traveler learns things at second-hand and by the halves, and is poor authority" (Thoreau 2004). Thoreau called those who travel not very wise, for all the wealth of knowledge about nature comes from observation while staying in one place (Thoreau 2004, 20–21). It can be presumed that the best place is where you live or a place close to home. It is there that one can discover the beauty of nature and there one develops respect for it. Thoreau repeatedly marveled at the beauty of nature with all its richness.

Thoreau admired the world of flora and fauna. The latter was an important area of observation, but it also influenced his moral stance, leading him to adopt vegetarianism and reject his beloved pastimes – hunting and fishing. Abandoning all forms of harm to animals allowed Thoreau to live in harmony with them. He tells, for example, of a local mouse that, over time, felt so confident in his company that it collected crumbs at his feet or nibbled on the cheese that Thoreau held in his hand (Thoreau 2004, 317). His house and surroundings were home not only to a mouse, but also to a lapwing and a thrush, and the documentation of his stay in the woods is full of colorful descriptions of the behavior of the animals he had the opportunity to observe. He often anthropomorphizes the scenes he watches, when he compares the ant and its brave behavior to Achilles wanting to rescue Patroclus and then avenge him, or when, during a race with an ant, he tries to guess the animal's thoughts (Thoreau 2004, 322–323).

An interesting strand of Thoreau's relationship with nature is his perception of it through the prism of Far Eastern spirituality or references to magnificent, wondrous places that do not always exist in the real world. In an experiment, Thoreau assumed that the "pure Walden water is mingled with the sacred water of the Ganges" (Thoreau 2004, 416). This is how the real world is described by Thoreau: through reference to the mythical world or religious beliefs. The same water that Thoreau goes for every morning is also fetched by a Brahmin servant who is a priest of Hinduism's main deities: Brahma, Vishnu, and Indra. Just as in the morning the philosopher at Lake Walden bathes his mind in the philosophy of the *Bhagavad Gita*, so this priest begins his day by reading the Vedas. The mystical language of Thoreau's descriptions of Walden Lake does not preclude a simultaneous fascination with research. Thoreau spent some time researching which living organisms inhabit the lake and was also concerned with the structure of the bottom of the body of water. The insightfulness of his mind can be evidenced by the fact that he was one of the few, perhaps the first, to accurately measure the lake's depth – the lake had previously been regarded as a bottomless reservoir. When an employee of Yale University's Osborn Zoological Laboratory, Edward S. Deevey, compared Thoreau's findings with contemporary measurements in 1939, they proved to be highly accurate (Cieplińska 2011, 296).

Thoreau was a very astute observer of nature and a true lover of it.

> We can never have enough of nature. We must be refreshed by the sight of inexhaustible vigor, vast and titanic features, the sea-coast with its wrecks, the wilderness with its living and its decaying trees, the thundercloud, and the rain which lasts three weeks and produces freshets.
>
> (Cieplińska 2011, 444)

He delighted even in those natural phenomena that arouse negative feelings in many people; although, for example, the sight of a vulture feeding on carrion was abhorrent to the philosopher, at the same time he saw it as an example of the scavenger drawing on life-giving power. He interpreted the sight of a dead horse as an expression of the richness of life of Nature, which can sacrifice some of its creatures to become food for others. In everything that happens, he saw the perfect harmony and organization of Nature, so that everything best serves its overall development.

4.2.4 Summary

Thoreau showed that a profit-centered civilization is not the path to moral perfection. The goal of his life was the pursuit of his own perfection. His path of becoming the best possible version of himself was a counterproposal to the emerging consumer civilization. He himself never wanted to accept the enslavement to which other Concord residents were subjected, which he felt was an obstacle to experiencing a high quality of life. Simplicity in outer life and the highest ideals in inner life – this seems to have been Thoreau's life motto. He saw others as mindless machines that fell into the cogs of earning and increasing wealth, thus condemning themselves to a life of imprisonment. Wealth for him was the abundance of virtue and moral excellence in which he could polish himself at Lake Walden.

The ecological ethos shown by Thoreau is based on the ideals of rejecting consumer civilization and worldly goods in favor of higher goods, character development, and contemplation. He professed to accept nature as it is and to accept life in harmony with it, despite all the inconveniences. What makes Thoreau's novel enticing is its authenticity: it is not a moralizing work that is an interpretation of exemplary moral behavior but a journey through the discovery of one's own experience of putting into practice the philosophy he professed. Thoreau inspired others to take action to protect the environment, but his love of nature was secondary, born of an appreciation of the value of living alone, which served to discover the best version of himself. Nature is valuable, but it is a backdrop for the key theme of *Walden; or, Life in the Woods* – his own journey toward moral excellence. This story of life in the woods cannot be reduced to a story of love for nature, for its essence is the individual experience of the journey through the meanders of perfecting moral character. Thoreau seemed to have been well prepared internally for this experience. Having settled in the woods, he had no great needs. Rejecting the comforts offered by the emerging consumer culture does not seem to have challenged him. He sees it as freedom from the shackles imposed by having an excess of things. *Walden; or, Life in the Woods* is a story about culture's rejection of consumerism in favor of the search for quality of life; it is a story about the fact that what is important is not outside but inside the moral agent. Finally, Thoreau's work is influenced by the *Bhagavad Gita*; it is a search for what gives liberation to the soul, rather than what binds it to the world. Against this background, nature appears as an ally of the hero's ascetic lifestyle, as a teacher of true life and humility, as the best place to become a better person, to be one who observes the essence of life rather than wasting time on mundane things.

Notes

1 At the time of Thoreau's life, the residents of Concord made their living mainly from agriculture and trade. It is a town in Massachusetts, founded in 1635, which became an important intellectual center centered around the persons of Ralph Waldo Emmerson and Thoreau. To this day, Thoreau's restored cottage is one of New England's greatest tourist attractions.

2 Cafaro points out that, before him, Thoreau's philosophical potential had been recognized by other philosophers who analyzed specific aspects of his thought, including Stanley Cavell (1972), who analyzed Thoreau's epistemology, and Bradley Dean and Laura Dassow Walls (1995), who traced Thoreau's contribution to the discussion of the relationship between philosophy and science. Bradley P. Dean organized hundreds of pages of Thoreau's manuscripts and contributed to their publication in the volume *Wild Fruits: Thoreau's Rediscovered Last Manuscript* (2000) and helped with the publication of *Faith in a Seed* (1996) and *Letters to a Spiritual Seeker* (2004).

3 The literature on the subject stresses that transcendentalism does not quite deserve to be called a philosophical system. Transcendentalism is more of a movement than a philosophy. It is a current that draws on philosophical inspiration, but it itself may not be fully called a philosophical system. Cieplińska points out that literary theorists even call it a gospel, an outburst of enthusiasm and a wave of feelings. It combines everything that goes beyond common sense (cf. H. Cieplińska 2011, p. 6).

4 New England Transcendentalism is a philosophical and literary current that was influenced by many different inspirations. It is stressed that classifying Thoreau as a representative of this current raises some objections. Furtak (2017), citing the research of Octavius Frothingham (1876) and Deborah Slicer (2013), points out that in the very history of Transcendentalism in New England, Thoreau is taken into account marginally, and his connection with this group of thinkers was not significant. Rather, it is pointed out that opposition to Emerson's concept of nature as a symbol was Thoreau's key concept, meaning that going beyond the ideas of transcendentalism was the essence of his philosophical views. Accordingly, it is rather suggested that Thoreau should be regarded as a representative of modern philosophy, independent of transcendentalism.

5 Thoreau's *Walden; or, Life in the Woods* was inspired by the *Bhagavad Gita*. His fascination with Hinduism and the teachings of dharma influenced his emphasis on the role of duty, the sacred destiny of man's birth.

6 Both admit that Thoreau influenced the formation of their political concepts (cf. P. Cafaro, Georgia 2004, p. 14–15).

7 Emerson is a figure of unique importance to American culture. He is often referred to as the American Socrates or the Sage of Concord. He is considered the first American thinker whose influence reached Europe. He is the father of transcendentalism. It is recognized that his figure was inspiring for not only philosophers but also poets, including Adam Mickiewicz (cf. Cieplińska 2011, p. 10).

8 Thoreau built the house in a minimalist manner, using wood from the forest and from a demolished house purchased for this purpose, with the kind help of friends (including Emerson, who also helped in this endeavor), thus setting an example of very rational management of scarce resources. The cost of building the house alone was only \$28.12½ (Thoreau 2006, p. 70–71). Of course, the purchasing power of the dollar was somewhat different at the beginning of the second half of the 19th century, but the house was nevertheless built very frugally. Despite this simplicity and modesty, the house and its construction have lived to see architectural analysis (cf. Barksdale 1999).

9 Thoreau points out locals' criticism of vegetarianism, himself pointing out the contradiction in the behavior of the farmer who criticizes a plant-based diet as insufficient to maintain proper bone structure. At the same time, the same farmer bases his work on the strength of oxen, which he herds to pull the plow despite their supposedly weak bones (cf. Thoreau 2006, 14). Thoreau's dietary minimalism

followed not only the rules of vegetarianism, but also principles derived, for example, from the Pythagorean school. The Pythagoreans did not eat beans because they believed that they consisted of the same decomposed matter from which man was created. Hence Thoreau too, inspired by the principles of the Pythagorean sect, rejected beans. However, he cultivated them with great pleasure (he even devoted a chapter of his book, *Walden; or, life in the woods*, to the description of a bean field) in order to exchange his harvest for rice.

10 *Dharma* is one of the key terms in Far Eastern philosophies and means, among other things, a set of ethical norms and duties that everyone must fulfill. These duties depend on one's age, status, professional group, social position, or situation.

11 The symbolic field of war between good and evil.

12 *Kalpa* – a unit of time, in Hinduism corresponding to one day of Brahma. A *kalpa* consists of one thousand mahayuga cycles, made up of four yugas (*satja, treta, dwapara,* and *kali*). A *kalpa* lasts 4.32 billion earth years. The Brahma night lasts the same amount.

13 An interesting analysis of Thoreau's perception in newspapers and magazines can be found in William J. Richard's book. His analysis shows the importance of Thoreau in American mainstream culture (2018).

14 Thoreau's book *Civil Disobedience* (1949) encapsulated his understanding of what it means to resist government action by breaking the law (reckoning with the consequences of such action). Most often, civil disobedience involved demonstrative, public nonviolent opposition to authority, and as such was the inspiration for Mahatma Gandhi. This attitude stemmed from disagreement with certain actions of the state. Thoreau, expressing his disapproval of the system of slavery, which he repeatedly criticized, refused to pay taxes. He believed that moral law was superior and more important than state law. The latter can always be rejected when it goes against moral principles.

15 Treanor points out that simplicity in Thoreau's terms has three dimensions and refers to the professional, material, and intellectual spheres. For a discussion of these three spheres, see the chapter on the narrative concept of EVE.

16 The scientific and technological revolution was received enthusiastically by most of the public, raising hopes that through the practical application of science it would be possible to eradicate diseases, increase the welfare and improve the working conditions of society, equalize social differences, and even solve social problems. This scientistic optimism was not shared by the Luddites. The Luddite movement arose in Britain in the late 18th and early 19th centuries. Its initiators were weavers and textile workers who had been rendered unemployed by the mechanization of this sector of the economy. The movement was named after General Ned Ludd, who was the leader of the movement, although there are doubts as to whether this figure existed. The essence of the movement was opposition to industrialization. Nowadays, neo-Luddites, following the example of the Romantics, reject technical achievements as the enemy of a well-lived life and the cause of human alienation (cf. Dusek 2011, 198–200).

17 He described such food preparation as poetic and contrasted it with cooking on a stove, which he described as chemical (cf. Thoreau 2006, 356).

18 Ascetic – not only in the sense of a hermit absorbed in meditation, but also in the sense of a person who reduces his needs to a minimum. A common myth accompanying analyses of this work by Thoreau is to attribute to him the image of a hermit, but this only partially captures the essence of his experience (cf. R. Schneider 1995, 192).

19 Thoreau likens the palm of the hand to a spread palm leaf (this is also a play on words, since the word for palm in English is palm) with lamellae and veins; the ear is a lichen (*umbilicaria*) on the side of the head, and the lips are petals or tongues protruding from the mouth (cf. R. Schneider 1995, p. 430–431).

20 Thoreau cultivated a small patch of land to gain the necessary resources for consumption or for exchange. Nevertheless, cultivation, according to Cafaro (Cafaro, Georgia 2004, p. 22), was a reference to shaping oneself, to cultivating one's character. Thoreau suggests that hard work refines us and physical labor is important, but – as pointed out earlier in the book – a person should not work more than necessary. Physical labor must not become a dirigible leading to the acquisition of more unnecessary goods. Thoreau himself regarded it as a tool for survival, and at the same time he paid attention to having time for scientific work. However, it is claimed that his approach to nature "was nurtured more by empathetic experience, imagination, and romantic poetry than by hard science" (cf. Brenner 1996, 129).
21 Thoreau here refers to a New Testament passage (Matthew 5:45): "He causes his sun to rise on the evil and the good and sends rain on the righteous and the unrighteous."

References

Andersen N., *Exemplars in Environmental Ethics: Taking Seriously the Lives of Thoreau, Leopold, Dillard and Abbey*, "Ethics, Place & Environment" 2010, vol. 13, no. 1, p. 43–55.

Bałżewska K., *Czarny wariant "Bildung": o relacjach między "Czarodziejską górą" Thomasa Manna a "Szpitalem Przemienienia" Stanisława Lema*, "Pamiętnik Literacki: czasopismo kwartalne poświęcone historii i krytyce literatury polskiej" 2012, vol. 33, no. 1, p. 111–128.

Barksdale W. M., *Thoreau's House at Walden*, "The Art Bulletin" 1999, vol. 57, no. 2, p. 303–325.

Bradley D. (ed.), *Wild Fruits: Thoreau's Rediscovered Last Manuscript*, New York 2000.

Brenner F. J., *Historical and Philosophical Development of Environmental Ethics*, "Bios" 1996, vol. 67, no. 3, p. 129–134.

Brook Farm, 1975, https://www.cityofboston.gov/images_documents/Brook%20 Farm%20%2314%20Study%20Report_tcm3-42379.pdf (Access: April 12, 2021).

Cafaro P., *Thoreau's Living Ethics. Walden and the Pursuit of Virtue*, Georgia 2004.

Cavell S., *The Senses of Walden*, Chicago 1972.

Cieplińska H., *Przedmowa*, in: H.D. Thoreau, *Walden, czyli życie w lesie*, tłum. H. Cieplińska, Poznan 2011.

Dusek V, *Introduction to the Philosophy of Technology*, transl. Z. Kasprzyk, Cracow 2011.

Frothingham O.B., *Transcendentalism in New England: A History*, New York 1876.

Furtak R. A., *Henry David Thoreau*, E.N. Zalta (ed.), in: *The Stanford Encyclopedia of Philosophy* (Fall 2017 edition) 2017, https://plato.stanford.edu/archives/fall2017/ entries/thoreau (Access: May 13, 2021).

Goodman R., *Transcendentalism*, E.N. Zalta (ed.), in: *The Stanford Encyclopedia of Philosophy* (Summer 2017 edition) 2017, https://plato.stanford.edu/archives/ sum2017/entries/transcendentalism (Access: April 1, 2021).

Gordon J., *Transcendental Ideas: Social Reform*, n.d. http://archive.vcu.edu/english/ engweb/transcendentalism/ideas/fruitlands.html (Access: April 12, 2021).

Hillway T., *The Personality of H.D. Thoreau*, "College English" 1945, vol. 6, no. 6, p. 328–330.

Mooney E. F., *Excursions with Thoreau. Philosophy, Poetry, Religion*, New York 2015.

Morawiec E., *Odkrycie egzystencjalnej wersji metafizyki klasycznej. Studium historyczno-analityczne*, Warsaw 2004.

Myerson J. (ed.), *The Cambridge Companion to Henry David Thoreau*, Cambridge 1995.

Saadi of Shiraz, *Gulistan or Rose Garden*, Boston 1865.

Sandler R., Review: *Thoreau's Living Ethics: "Walden" and the Pursuit of Virtue, Philip Cafaro*, "Environmental Values" 2006, vol. 15, no. 1, p. 135–138.

Schneider R., *Walden*, in: *The Cambridge Companion to Henry David Thoreau*, J. Myerson (ed.), Cambridge 1995, p. 92–106.

Slicer D., *Thoreau's Evanescence*, "Philosophy and Literature" 2013, vol. 37, p. 179–198.

Stoller L., *Thoreau's Doctrine of Simplicity*, "The New England Quarterly" 1956, vol. 29, no. 4, p. 443–461.

Thoreau H. D., *Civil disobedience*, New Jersey 1949.

Thoreau H. D., *Walden; Or, Life in the Woods*, "The Pennsylvania State University" (for the source electronic book file version) 2006.

Walls L. D., *Seeing New Worlds: Henry David Thoreau and Nineteenth-Century Natural Science*, Madison, Wisconsin 1995.

William Richard W. J., *Finding Thoreau: The Meaning of Nature in the Making of an Environmental Icon*, Boston 2018.

5 Naturalistic, teleological, and pluralistic environmental virtue ethics

The author of the naturalistic, teleological, and pluralistic concept of EVE is Ronald Sandler.[1] He presented his concept in 2007 in the publication *Character and Environment. A Virtue-Oriented Approach to Environmental Ethics* (2007), which is one of the most important works in environmental virtue ethics. According to Cafaro (2010), it was the best theoretically refined work on EVE that had been published by 2010, and I can state that it is still the best in 2023. Its strength is its solid theoretical background. Sandler points out that environmental ethicists often make little reference to contemporary works on virtue ethics (Sandler, *An Interview with*). Hence, he refers to important theories in the field and analyzes them from an environmental perspective, his main inspiration being the views of Margaret Anscombe, Rosalind Hursthouse, and Philippa Foot.[2] This type of approach makes Sandler's concept of EVE an important contribution to the theory of virtue ethics (cf. Calder 2010, 234). Sandler often uses the phrase "virtue-oriented ethics" in place of "virtue ethics."[3] This term is analogous to the concept of "virtue theory" that was defined in the first part of this monograph, and in this case it means aretological reflection in relation to environmental ethics. This rhetorical device in Sandler's writing helps to avoid theoretical disputes around the understanding of virtue ethics.

5.1 The foundation of naturalistic, teleological, and pluralistic environmental virtue ethics

Sandler emphasizes that what

> makes an environmental ethic virtue-oriented, then, is that it evaluates our relationships, actions, practices, and policies regarding the environment in terms of virtues and vices, and that it emphasizes cultivating environmental virtue – i.e., those character traits that are conducive to promoting human and nonhuman flourishing – in addition to performing particular actions.
>
> (undated interview with Sandler, 2023)

DOI: 10.4324/9781003433156-8

He refers to the concept of moral character but seeks so-called environmental character. In his view, proper environmental ethics is character ethics that provides guidance on the qualities and dispositions that a moral agent should have in relation to the environment (Sandler 2005a, 2). Sandler transforms the ethical question "how should we live?" into the question "what kind of person should we be?", which refers to Hill and his considerations captured in the phrase "What sort of person would destroy the natural environment – or even see its value solely in cost/benefit terms?" (Hill 1983, 211).

Sandler's concept is an interesting example of combining the themes of classical virtue ethics with modern currents and relating them to the issue of humans' relationship to their natural and social environment. The influence of modern virtue ethics can be seen in the naturalistic nature of EVE, which was inspired by the concepts of two ethicists, namely Rosalind Hursthouse and Philippa Foot. In turn, the reference to classical virtue ethics is expressed in the teleological nature of Sandler's concept. Aristotle's virtue ethics inspired this philosopher to view moral fitness as leading human beings to happiness and fulfillment in life, and to see that human actions are not limited to the realization of deeds leading to eudaimonia, but alongside these there are also several non-eudaimonic goals that are important to the moral agent. This extension of eudaimonia to include non-eudaimonic goals led Sandler's ethics to be called pluralistic.

5.1.1 *The naturalistic nature of environmental virtue ethics*

The starting point for Sandler's naturalistic conception of ethics is to emphasize the importance of humans' biological nature, which from the perspective of virtue ethics can be seen as troublesome. The influence of Christian thought on philosophy has contributed to the appropriation of virtues for higher purposes, while considering the needs of the body as secondary. This means that corporeality and the needs of the body are excluded from the perspective of virtue. Sandler challenges this approach: he believes that the bodily element is central to environmental virtue ethics, since the body is what connects us to the natural world. Sandler emphasizes that the connection between humans and the rest of the natural environment is precisely our biological nature. In line with this view, he points out that we are subject to the same evolutionary processes as every other species on Earth, and we are also made of the same kind of matter as every other living organism (Sandler 2007, 68).

This view is rooted in the ethical naturalism represented by Philippa Foot (2001) and Rosalind Hursthouse (1999). Hursthouse points out that in discussions about ethically good people we have not suddenly begun to use the term "good" in an entirely new "moral" meaning, but in a sense that has its origin in the colloquial understanding of the word (Hursthouse 1999, 175). Botanical or ethological assessments of living things use the terms "good" or "defective" to refer to specimens of a species in order to evaluate their body parts or behavior.

This carries over to the evaluation of moral entities and their characters as ethically good or bad. Hursthouse also points out that we perceive ourselves as material objects that exist in a particular space, rather than as immortal objects with immortal souls or as entities that are persons or rational beings. Ethical and nonethical analyses are in some ways analogous to those known in biology (Hursthouse 1999). Scientific naturalism provides a definite framework for evaluating the goodness of an animate thing as an animate material thing, while at the same time it contains a rationale for evaluating humans as objects of a given class of being. Belonging to the world of animate objects regardless of species (human, animal, or plant) is governed by certain rules or, as Philippa Foot puts it, a certain "grammar" (Philippa Foot 2001).

Ronald Sandler makes this naturalistic conception of the good the starting point for his reflections on the nature of virtue. Following Philippa Foot and Rosalind Hursthouse, he emphasizes that belonging to a particular species and genus contributes to adopting the given definition of the good that is inherent in this life form. When it comes to plants, the good is that their parts (leaves, roots, flowers) and actions (absorbing water, flowering, shedding leaves) serve the survival of their species. Weak roots or disruption of any of the vital functions leads to the weakening and defective development of the plant and is a denial of the natural good inherent in plants of this species. When it comes to more-organized forms of life, what constitutes good usually means something more than concern for survival and ensuring the continuation of the species. "For species whose members are sentient, there is avoidance of pain and experience of pleasure. For species whose members are social, there is being a member of a well-functioning social group" (Sandler 2007, 15). Following this line, Hursthouse formulates four goals pursued by a good social animal: 1. survival; 2. ensuring the continuity of the species; 3. species-specific freedom from pain and species-typical entertainment; and 4. functioning well in a social group (Hursthouse 1999, 202). The assessment of whether an individual is good involves the following areas: parts, functioning, action, desires, and emotions. In biological terms, the fulfillment of these conditions in the preceding areas is important for a particular individual to be considered as good.

Such an assessment is not without its difficulties as any assessment should always take into account the individual living conditions of the individual being analyzed. It may turn out that a particular individual is good, even though it does not achieve all the goals listed by Hursthouse. Sandler, to illustrate this, gives the example of a panther that cannot find a mate because all other panthers in the area have been killed, thus it cannot further the species (2007, 16). The lack of offspring in this case is the result not of a biological defect but of an unfortunate combination of events. Thus, from a biological point of view, an individual can be seen as good, even though it has not met one of the criteria established in the assessment.

Proponents of the naturalistic conception of the good extend it to the evaluation of human life. They emphasize that a good life consists of the realization of biological goals. Thus, the naturalistic position would assume the following formula for evaluating humans' natural goodness:

a human being is ethically good (i.e., virtuous) in so far as she is well fitted with respect to her (i) emotions, (ii) desires, and (iii) actions (from reason and inclination); whether she is thus well fitted is determined by whether these aspects well serve (1) her survival, (2) the continuance of the species, (3) her characteristic freedom from pain and characteristic enjoyment, and (4) the good functioning of her social groups – in the way characteristic of human beings.

(2007, 17)

The biological dimension of man's life is part of his good life and an essential element of it; however, unlike in the plant world, an assessment of this sphere does not completely capture the essence of man. Sandler wants to preserve this approach but stresses that it does not provide sufficient justification for ethics. After all, in addition to biological functions, a feature of humans is rationality, which distinguishes us from the rest of the natural world. Rationality is crucial in determining virtues. At the same time, the biological dimension of the human species should not be forgotten as it is the bridge between human and the natural world. The determination of what is a virtue depends on how specific qualities contribute to the full development of man as a representative of the species. More specifically, it involves defining the qualities that make a person the best version of himself and a good representative of the species (Sandler 2005, 5). It seems that emphasizing the biological dimension of humanity is important for capturing humans' relationship with nature. In addition, it makes it possible to "disenchant" conceptions of man that overemphasize the spiritual elements of his nature. Reference to biological nature can help find Aristotle's golden mean in virtue. However, it is necessary to be reasonable in the evaluation of the moral agent, that is, the evaluation must not be narrowed down to an analysis of only one of the spheres because then it becomes excessively reductionist.

Foot and Hursthouse point out that the way a person pursues these life goals is not solely determined by biology. It is influenced by rationality and is culturally determined – it goes beyond biological determinants (cf. Sandler 2007, 36). Human beings can even give up the realization of certain biological goals in favor of higher goals, and this in no way affects the definition of the good. There can be many examples here, such as a scientist or environmental activist giving up having offspring to serve the world, or an ascetic consciously limiting the satisfaction of bodily needs to pursue religious goals. In such a case, giving up biological goals does not diminish the quality of life; on the contrary, it makes life a path to higher goals.

Sandler stresses that scientific naturalism does not provide sufficient justification for ethics. The rationality that characterizes humans also makes them pursue their biological goals differently than animals. At the same time, the goals that man sets for himself if he wants to achieve full development are different from those of nonhuman animals. The naturalistic approach is the starting point for further consideration:

> we cannot look only at our evolutionary history or the biological functions of our parts or systems to settle the ends constitutive of human flourishing. We must also consider, critically of course, common beliefs about what constitutes human flourishing, as well as the ways rationality, culture, and technology shape and provide novel possibilities for our life form.
>
> (Sandler 2007, 22)

Thus, defining the goals relevant to full human flourishing requires considering evolutionary development, biological data, and the prospects that rationality opens for humans.

According to Sandler, one can see the influence of rationality on human development in at least three spheres: meaningfulness, knowledge, and autonomy. He stresses that we are predisposed to seek the meaningfulness of our lives in a way that nonrational animals cannot. Moreover, certain values – Sandler mentions courage, perseverance, and optimism here – are virtues because they help humans realize the meaningfulness of life. The biological view does not fully capture this intangible human goal of striving to give meaningfulness to one's existence. The mere fulfillment of goals derived from the naturalistic assumption also does not result in the satisfaction of man's aspiration to give his own life meaningfulness. The second area affected by rationality is knowledge, the accumulation of which at the individual and supra-individual levels is an important area of human life that does not only arise from biology but also contributes to the creation of culture. "The accumulation and transmission of acquired knowledge between people and over generations is among the most striking and distinctive features of the way human beings go about the world" (Sandler 2007, 24). A system of educational and research institutions has grown up around the process of knowledge creation. In addition, continuous learning, lifelong improvement, and the ability to expand knowledge have recently been described as some of the most important characteristics of modern man. The ability to create new knowledge exceeds the ability to assimilate and apply it. Knowledge, according to Hans Jonas (2010), leads to revolutionary change, and its dynamic development is associated with rapid change. The measure of scientific progress is humans' ability to enjoy the fruits that this development brings. Thus, as Jonas argues, when a person in adulthood has to ask his children or grandchildren to explain reality because he himself does not feel equipped with sufficient knowledge, we can speak of revolutionary progress (2010, 46). It follows that simple forms of knowledge creation and certain behaviors resulting from this in the animal world cannot match the level of accumulation and creation of new knowledge in the human world. The third key element of humanity is autonomy, which means that the moral agent has many ways of pursuing his goals; he is not limited by a specific scenario of action and behavior that leads to full personal development. Autonomy is used to choose the best project that will give meaning to the individual's life and to help decide how to pursue the goals necessary for full personal development, considering the opportunities available to the individual.

Thus, meaningfulness, knowledge, and autonomy complement humans' biology; all three constitute humans' good life and are the basis for setting goals relevant to the realization of full personal potential. In summary, in the case of human beings as a biological and cultural creation, biological functions and social functioning are crucial, and finally full human flourishing is determined by rationality. Biology is the foundation on which humans' functioning is based, but life goals, as well as the full realization of their own potential, are not limited to the sphere determined by it. Taking these arguments into account, Sandler proposes the following formulation of goodness in the ethical sense:

> a human being is ethically good (i.e., virtuous) in so far as she is well fitted with respect to her (i) emotions, (ii) desires and (iii) actions (from reason and inclination); whether she is thus well fitted is determined by whether these aspects well serve (1) her survival (2) the continuance of the species, (3) her characteristic freedom from pain and characteristic enjoyment, (4) the good functioning of her social groups, (5) her autonomy, (6) the accumulation of knowledge, and (7) a meaningful life – in the way characteristic of human beings (i.e., in a way that can rightly be seen as good).
>
> (Jonas 2010, 25)

The naturalistic part of virtue ethics requires some comments considering the impact of rationality on the transformation of the environment, including human biology. Advances in knowledge mean that we can influence the cells, genes, and nervous system of living organisms. We can replace diseased tissues or organs with healthy ones, and we can also stimulate their process of returning to full function by means of technical devices. Thus, the perspective of biological evaluation adopted by Sandler due to the belonging of the evaluated moral agent to a particular life form becomes an area of variability in the modern world. Thanks to rationality, humans are able to transcend biological conditions in many cases (cf. Sandler 2014).

What's more, research and bold promises to improve human capabilities are beginning to enter the realm of human rationality, offering the hope of enhanced human cognitive abilities. Equally bold proposals for improvements in human biology are being made as part of transhumanism projects, some of which include developing moral character by stimulating the development of empathy.

When some of these projects become a reality, and some of the improvements in the way humans function are put into practice, we will not be able to speak of a strictly naturalistic approach. However, as Sandler emphasizes, these projects are, for now, a matter of the near or distant future (2007, 26). Scientific developments may lead us to change our views not only on the discussed naturalistic approach; they can contribute to a change in our perception of other human goals, as well as to the formulation of new goals that are more adapted to the conditions of human life. Meanwhile, the goals of a good human life are still influenced by our biology and are in line with the naturalistic view.

5.1.2   The teleological nature of environmental virtue ethics

The essence of teleological ethics is the eudaimonistic character of virtue ethics, which states that the goal of human life is happiness, understood as fulfillment in life. In eudaimonistic ethics, fundamental questions include "how should I live?" or "what kind of person should I be?" The latter question is about the character traits that allow one to be a good person. Virtues and moral character are tools that lead to the realization of the goal of a fulfilled, happy life. A central tenet of eudaimonistic ethics is the claim that good character is valuable not only because it leads to good actions, but also because it benefits those who possess it. The claim that the moral agent benefits from possessing virtues and a good character is the reason for the charge leveled against virtue ethics, namely that of selfishness (cf. Annas, 1993; Hursthouse 1999; Hurka 2001). The literature emphasizes that virtues, including environmental virtues, bring many benefits to the moral agent. Appreciation, respect, admiration and love of nature help people derive satisfaction from being in nature and enjoy their relationship with it. Good environmental character is an element of full personal development, and those who possess virtue are always morally superior to those who do not. Virtue is so valuable and significant that it enriches the one who possesses it, giving him additional value. Sandler (2007, 2), following John Muir, emphasizes that everyone needs places to play and pray where nature can heal and strengthen the body. In addition, the moral agent has the ability to discern what serves his flourishing. According to Sandler, it is precisely the ability to discern what is good for the moral agent that is one of the skills constituting his striving for human excellence (Sandler 2007, 31).

An important role in Sandler's view is played by the connection between virtues and human excellence. Virtue, which constitutes happiness and human excellence, is necessary for a human being to bring out his best qualities and become the best version of himself. This is because virtues help the moral agent correctly perceive what is good and right, thus guiding his conduct so that it is virtuous (Sandler 2007, 30). A virtuous person knows how to realize the virtues in the right way, even if this requires certain sacrifices or renunciations. A moral agent knows how to direct his aspirations to avoid suffering and fulfill his proper desires. He knows how to recognize which action, goal, or thing is good and serves his development. Hence, virtue is a key ingredient of achieving human excellence; it constitutes it but at the same time allows one to recognize what is necessary to ensure it.

Thus, the good in human life is the realization of biological goals, as well as the goals of human excellence and happiness. These goals are determined by biology (in a manner characteristic of a rational being), but at the same time they are pluralistic goals (as will be shown in the next section). The concept of an entity that pursues its goals according to its naturalistic conditions is somewhat similar to following Paul Taylor's ideas. According to Taylor, every animate entity is the teleological center of life, by which he means that everyone pursues the typical goals and goods of that animate form. A being is a goal-oriented entity, and its pursuits are determined by its form of life (Taylor 1986, 221–124). This view is also a naturalistic view as, for Taylor, every animate organism has its own goals and aspirations that arise from its biological conditions, even when they are limited to survival, preservation of health, and reproduction.

Although there are approaches in virtue ethics that are non-teleological in nature, Sandler rejects them. In this case, his views are very consistent with his criticism of so-called ecological heroes. He stresses that we cannot adjudicate the virtues or virtuous character of a moral agent on the basis of merely observing what this individual is or was. Merely showing that certain qualities are praiseworthy or appropriate is not an argument for nurturing them or recognizing them as virtues. The demonstration of virtues must be accompanied by reflection on whether these qualities allow us to achieve eudaimonious goals. Evaluating a trait as having certain desirable inner qualities or a beautiful interior is not appropriate as it is inadequate to reality and too hasty. A given trait should be evaluated not on the basis of beliefs about the inner beauty it guarantees, but on the basis of whether it leads to good for the moral agent, whether it ensures happiness and human excellence, and whether it influences the realization of non-eudaimonistic goals that are important to the moral agent.

5.1.3 *The pluralistic nature of environmental virtue ethics*

The pluralistic nature of Sandler's EVE can be seen in two areas: first, by pursuing non-eudaimonistic goals that are part of the good life; second, in the pluralistic approach to defining the entities considered in moral reflection. In both cases, there is a move away from narrowing ethics to a more comprehensive ethical outlook. Character evaluation ethics notes that a virtuous person will find fulfillment not only in activities that serve his personal benefit. According to Sandler, a person with the virtue of compassion has a natural predisposition to see the suffering of others and the desire to help them (2007, 20). He or she does not tend to calculate whether providing help will bring him or her any benefit, for seeing the suffering of other triggers the reflex to help in such a person. This means that such a person is also open in his life to the realization of non-eudaimonistic goals, which also constitute an element of an ethically good life. This approach can be called pluralistic.

Personal development and the pursuit of human excellence are important, but moral character is determined not only by eudaimonistic goals as some goals are not directly linked to the happiness of the moral agent. Thus, in addition to goals related to the moral agent, there are those that do not relate to himself, such as helping others. A proponent of the eudaimonistic concept would describe goals as eudaimonistic and would postulate that they lead a moral agent to human excellence. However, Sandler stresses that, in the pluralist view, non-eudaimonistic goals are important alongside eudaimonistic goals. An example of this is beneficence toward others, which is understood as a quality that in itself promotes the good of not the moral agent but another person. Such a view constitutes the pluralistic character of virtue, the essence of which is to serve the eudaimonistic and non-eudaimonistic goals of the moral agent.

Pluralism is also expressed in broadening the scope of moral reflection. Sandler, as he himself points out when he writes about how a good person should function well in a social group, is thinking about not only society but also the natural environment. Following Aldo Leopold, he refers to the idea of a biotic community. However, this does not mean narrowing the area of moral reflection to a theoretical position called holism (Sandler 2005a, 5). In considering the nature of environmental virtues, Sandler asks a question that is often raised in environmental ethics about the limits of moral reflection. Traditionally, all ethics is anthropocentric, but anthropocentrism in environmental ethics has become a kind of burden and even sometimes a serious charge. This goes along with the statement that ethics should also take into account other beings, not only human beings. However, the question also arises whether anthropocentric ethics can really protect nonhuman beings to the same extent as it cares for humans. Along with this doubt has come a kind of contestation of the anthropocentric tradition, with many ethicists facing accusations of species chauvinism. However, the discussion boils down to the basic question of who should be included in ethical consideration. Who should we be concerned with in this matter, and who can we leave out? This is the question of the boundary that marks our interest in a particular subject. Among representatives of environmental ethics, there is agreement on only one thing: ethics should protect not only human beings but also nonhuman entities. However, this leads to a dispute, the main directions of which can be reduced to the following questions: how far does the scope of moral reflection extend? What are the criteria for inclusion in protection? What priorities should we adopt in determining the importance of being in ethics?

The problem relates to the issue of moral considerability, which is increasingly emerging in ethics not only in the context of the environment but also in the face of new challenges in a globalizing world. Giving moral credit, according to Thomas Birch, can be explained by the following formula:

> to give moral consideration to *X* is to consider *X* (to attend to, to look at, to think about, where appropriate to sympathize or empathize with *X*, etc.) with the goal of discovering what, if any, direct ethical obligations one has to *X*.
>
> (Birch, 1993, 315)

The key here is to point to what Birch calls "the club of (moral) *consideranda*," and this in turn leads us to the attributes associated with recognizing membership in this club (Birch, 1993). When we talk about moral recognition, we must remember that there are those who are in the "club" and those who are outside it. Many times, the framework of the club has been expanded and the criteria for belonging to it have been changed. During slavery, outside the "club" were servants, who were deprived of many rights; for most of the history of Western culture, women did not have the right to vote, to education (either not at all, or not at all levels and in every direction), to practice many professions, and in some periods were completely excluded from taking part in public life.

If we assume that some beings deserve moral recognition and some do not, we consequently adopt a certain criterion or criteria related to membership of the moral recognition club. For many environmental ethicists, life is the most prominent category that represents a kind of entry ticket to the moral recognition club. Biocentric ethics, such as that of Paul Taylor, recognizes life as a value in itself and extends ethical reflection to all life-bearing entities. Every animate being deserves protection by virtue of the very fact that it is alive. However, determining whether an object is alive is sometimes controversial, at least in some cases; this is well illustrated, for example, in bioethical discussions concerning the beginning, duration, and end of human life. But the nonhuman world is not free of dilemmas in this regard either, as can be seen from the discussion on whether viruses can be considered animate.

This problem is contained in another condition formulated by Birch, which says that we should be able in a rational and nonarbitrary way to identify the trait(s) for membership of the moral recognition club. The problem of being able to properly identify certain qualities that can be considered a criterion for inclusion in the club of moral recognition was framed by Sandler in terms of the so-called epistemological gray zone. The difficulty in the precise definition of the criteria for membership of the club makes it difficult for us to qualify a certain entity as belonging to a certain category, or to unambiguously determine whether it meets the criteria for moral consideration (cf. Sandler 2007, 40). Despite this difficulty, we should do everything possible to protect the entities we have defined as having qualities that include them in the club of moral consideration. In this view, the qualities that qualify an entity as belonging to the group of morally recognized entities give rise to an obligation toward them.

Therefore, Birch's next condition is that we should establish practices that enhance the qualities that define membership in the club and ensure the integrity of the club, while taking care to maximize the well-being of its members (cf. Birch, 1993, 315).

The capabilities approach reveals many of the problems that ethics deals with. First, there is a group of entities that are included in ethical consideration, and there is a group of those that fall outside the scope of its interest. A form of exclusion from ethical decisions is apparent which can be considered controversial. At the root of this exclusion is the adoption of certain assumptions about the characteristics that an entity must have to be relevant to ethics and to be considered in moral choices. The moral consideration club could be presented in a concentric form that emanates in an increasing range from an individual person. Moral consideration could apply only to the moral agent (egoism), his family (nepotism), the inhabitants of the same country (patriotism), all of humanity (anthropocentrism), all sentient beings (pathocentrism), all living entities (biocentrism), or ecosystems (ecocentrism) (Sandler 2007, 40). Such a concentric view shows the expansion of the moral circle from the individual to ecosystems, from the moral agent to a larger reality that transcends the individual entity. This perspective is quite functional as it accurately shows the scope of moral reflection in a given ethic. It is a good illustration of the moral progress that has enabled humanity to move beyond concern only for its own species. Sandler notes that representation of the framework of ethics by means of concentric circles does not correspond with our life experience (Sandler 2007). This is because Sandler accepts this notation as a kind of form of reductionist framing that reduces every entity to a certain class and thus assigns it to a certain circle, enclosing it within it. Moreover, these circles can overlap; thus, some entities fall into a few categories.

The answer to this dilemma is the pluralist account proposed previously. Pluralism in this case implies that there is no single set of criteria for classifying a given entity as a member of the club of moral recognition. There is no single characteristic or set of characteristics that allows some to be included in our ethical reflection and others to be excluded from it. In addition, the pluralist approach corresponds better to our everyday ethical experience, in which we do not analyze what type of object a given entity is before taking action, but we can feel an obligation to any category of natural objects.

This demonstration of the non-eudaimonistic and teleological nature of certain traits that can be considered virtues forces Sandler to reformulate the conditions for determining what makes a character trait a virtue; thus, he proposes the following formula:

> a human being is ethically good (i.e., virtuous) in so far as she is well fitted with respect to her (i) emotions, (ii) desires and (iii) actions (from reason and inclination); whether she is thus well fitted is determined by whether these aspects well serve (1) her survival (2) the continuance of the species, (3) her characteristic freedom from pain and characteristic enjoyment, (4)

the good functioning of her social groups, (5) her autonomy, (6) the accumulation of knowledge, (7) a meaningful life, and (8) the realization of any non-eudaimonistic ends (grounded in non-eudaimonistic goods or values) – in the way characteristic of human beings (i.e., in a way that can rightly be seen as good).

(2007, 28)

Thus, a particular character trait can be considered a virtue when it serves both eudaimonistic and non-eudaimonistic goals regarding the dependent and independent goods and values of the moral agent. Such a very general formula applies to many spheres of human life. Unlike some interpretations of environmental virtue ethics, this concept does not overlook the social aspect and its role in human morality (cf. Rolston III, 2005).

5.1.4 Summary

As the preceding discussion shows, a character trait can be considered a virtue if it can be described as naturalistic, pluralistic, or teleological. Ethical goodness is naturalistic because it is grounded in scientific naturalism and considers the biological conditions of human existence. It is also pluralistic because it presupposes eudaimonistic and non-eudaimonistic goals, that is, those relating to the moral agent as well as those that are not directly related to it. Goodness is teleological, since character traits are evaluated in terms of their potential to facilitate the achievement of certain goals (Sandler 2007, 28). These considerations determine the meaning of ethical goodness. In Sandler's terms, the given understanding is the basis for determining what an environmental virtue is. Hence, in the next section of the book, I will take a closer look at Ronald Sandler's understanding of virtue.

5.2 Virtues in naturalistic, teleological, and pluralistic environmental virtue ethics

Ronald Sandler understands virtue as a character trait (cf. Sandler 2005b, 38). This understanding is grounded in the etymology of the term "morality." As MacIntyre points out, in neither Latin nor ancient Greek is there an equivalent word for "morality." The term etymologically comes from *moralis*, which literally means "pertaining to character." "where a man's character is nothing other than his set dispositions to behave systematically in one way rather than another, to lead one particular kind of life" (2007). The understanding of virtue as a character trait is deeply rooted in our culture. Thus, environmental virtues, in Sandler's terms, are the character traits of a person that make us take into account interactions and relationships with the environment (2005, 3). An environmentally virtuous person is equipped to take responsibility for the environment. A virtuous person has the right disposition to recognize what is right and perform it with the right motives (2005, 6).

5.2.1 What are environmental virtues?

Sandler views virtue from a naturalistic, teleological, and pluralistic perspective. Virtue for him is not only a moral ideal or an exemplary character trait; for him, it is also that which arises from our biology and that which serves eudaimonic and non-eudaimonic goals. In addition to biology, the improvement of the moral agent is crucial. Environmental virtue ethics in this view leads to the achievement of eudaimonia. Sandler, like many other environmental virtue ethicists, does not provide ready-made prescriptions and solutions. He does not claim that moral agent X should possess certain character traits in order to be said to be a virtuous person. He proposes certain goals set by biology, as well as non-biological factors, which are a kind of map in human striving. However, following this map is not a categorical imperative; it is rather a clue as to the areas where achievements can bring a person satisfaction, happiness, and even human excellence. Unlike plants, for humans there is no single scenario that will make them full-fledged representatives of their species. Indeed, it is easier to say, for example, that a rose flower has fully developed than to assess whether a person has achieved the goal of his life or has become the best version of himself.

Although aspirations to moral perfection can also be found in Thoreau's thought, Thoreau's views pose challenges to humans that an ordinary moral agent would not face. The harsh living conditions he imposed on himself made the ethos that follows from his concept an ascetic one. The standard of living at Lake Walden not only differed from the living conditions of the average representative of Western society in our time, but it was harsh even for his contemporaries. It is an example of a supererogatory ethos that only a handful of people can live up to. While Sandler emphasizes the importance of Aristotle's middle way, he sees the pursuit of moral excellence in this Stagirite's approach.

Sandler points out that, depending on the context, different environmental virtues take on significance. For example, among the virtues leading to environmental sustainability, he lists temperance, simplicity, farsightedness, attunement, and humility. Among the virtues expressing respect for nature, he sees care, compassion, non-maleficence, restitutive justice, and ecological sensitivity. Communion with nature, wonder, openness, attentiveness, aesthetic sensibility, and love are key (undated interview with Sandler). Sandler stressed that environmental virtues are many, and each can take different forms in particular situations. This multiplicity of virtues, as noted earlier, corresponds to the richness of the challenges posed by human interaction with the natural environment. At the same time, because of the multiplicity of virtues, there is no one-size-fits-all scenario for being virtuous in relation to the environment.

According to Sandler, the environmental dimension of a virtue is determined by its sensitivity to the environment. An environmental virtue, then, is any virtue that involves responsiveness to environmental entities. This is the type of virtue Sandler refers to as an environmentally responsive virtue (2007, 42). Another category is environmentally justifiable virtue, namely virtue that is "in part justified by environmental considerations (e.g., the worth of living organisms, beneficial relationships with environmental entities, or environmental resources)" (2007, 43). An environmentally productive virtue promotes and sustains environmental goods or values (e.g., ecological integrity or the flourishing of living organisms). Thus, "environmentally justified virtues, environmentally responsive virtues, and environmentally productive virtues are each a (not mutually exclusive) type of environmental virtue" (Sandler 2013).

Virtues also presuppose certain desirable dispositions necessary for their realization. They can be a source of the sensitivity and wisdom necessary to determine whether a virtue should be realized, and if so, how. Wisdom can help in making the right decision in a situation of moral conflict (Sandler 2018, 231). Hence, because of the fundamental importance of virtues in choosing appropriate actions, some philosophers give primacy to character ethics over ethics focused on moral evaluation of a deed. In environmental ethics, too, the importance of virtues is significant. They do not play a merely instrumental role as dispositions to act appropriately, but they contribute to the development of the moral agent and provide him with opportunities to improve his choice of the best possible actions.

According to Sandler, each person approaches the question of virtue differently in the same or analogous situation. To demonstrate this, he cites the words of Aristotle, who considers how a man who is cowardly, audacious, or brave acts. According to this Greek philosopher,

> the coward, the rash man, and the brave man, then, are concerned with the same objects but are differently disposed towards them; for the first two exceed and fall short, while the third holds in the middle, which is the right position: rash men are precipitate and wish for dangers beforehand but draw back when they face them, while brave men are keen in the moment of action, but quiet beforehand.
>
> (Aristotle 2007, 45)

This example shows how, in the context of the principle of the golden mean, the behavior of a moral agent differs depending on the presence of virtue or vice in that person. Every moral agent is driven by certain emotions, desires, or motives. Some of these flow from our biology. According to Sandler, this is a key factor in the process of determining virtues. As this philosopher notes, "discussion of what makes a character trait a virtue or vice begins from the naturalistic premise that human beings are essentially biological beings" (2007, 13). Our bodies are composed of matter, while we ourselves are subject to the laws of nature and determined by our genetic endowment. The biological factor is therefore the basis of our existence, and we cannot deny it without hypocritical reality. Like all other entities found in nature, we are determined by our biology, but in a different way. Our way of existence and functioning in the world is different from the way any other species exist and function in the world. However, this statement is not, in Sandler's case, an expression of species chauvinism, for he immediately adds that while we are different from other species, "being unique is nothing unique to us" (2007). Individuals of each species are different from individuals of other species, and humans do not have some special place in the chain of being. We have some unique abilities, but also in the natural world there are species that have abilities that we do not have (2007).

Moral pluralism is also expressed in the concept of virtues, of which there are many and which manifest themselves in different ways. In moral pluralism, the scope of individual virtues is determined by the bases "to which a virtue (or, more accurately, the person who possess the virtue) is responsive" (2007, 42). For example, the basis of compassion is the suffering of others, and the form of responsiveness is caring for others – wanting to help and acting to reduce the suffering of another (2007). Both the extent and manner of responsiveness to a particular situation are determined by the nature of the moral virtue. As already mentioned, virtue is a character trait that helps the moral agent to recognize his obligation and behave in a virtuous manner; at the same time, it is adequate to the situation at hand. A virtuous person knows how to read the context of a situation that requires a response correctly enough to know which virtue is relevant in a given situation and to what extent it should be applied. Ronald Sandler's approach assumes a teleological conception of virtue.

5.2.2 Strategies for specifying environmental virtues

The concept of virtue is based on the assumptions of Sandler's naturalistic, teleological, and pluralistic virtue ethics. It refers to the biological nature of human beings; it meets the criterion of the naturalistic conception of ethics and is subject to teleological assumptions in the pursuit of eudaimonistic and non-eudaimonistic goals. In addition, the catalogue of virtues and ways of pursuing them are very extensive, which is an example of pluralism. This abundance of environmental virtues and vices is no simplification when it comes to identifying who possesses the character that enables protection of the environment and what qualities/vices can be defined as environmental. Sandler specifies in this regard that an "environmentally virtuous person is disposed to respond – both emotionally and through action – to the environment and the non-human individuals (whether inanimate, living, or conscious) that populate it in an excellent or fine way" (2005a, 3). Sandler describes this definition as formal, while noting that it does not provide any substantive description of a person who is virtuous toward the environment. This makes it difficult to discern who possesses character described as environmental and which virtue is environmental.

According to Sandler, there are several strategies for specifying a particular trait as an environmental virtue (2005a, 4–6). The first way involves looking at interpersonal virtues and trying to extend the standard virtues of this kind. Each virtue is normative for a certain range of things, actions or interactions, and this range sets the framework for its application. As examples, Sandler writes "the field of honesty is revealing or withholding of truth; the field of temperance is bodily pleasures and pains; and the field of generosity is the giving and withholding of material goods" (2005a, 4). This approach involves broadening the scope of specific virtues to see if they can at least partially relate to the environmental context. An example of this is compassion, which, as a virtue, can apply to another person or can relate to nonpersonal entities.

For example, the virtues of care, helpfulness, environmental responsibility, and many others relate to interpersonal relationships, such as care for the environment, which, in Hans Jonas' view, is most exemplarily expressed in parental concern for a child, in which the child's naked ontic "to be" enforces the parents' duty to the child (Jonas 1996, 186). The same is true, according to Sandler, of the virtue of friendship. Friendship toward nature is an extension of the same virtue, which is a relationship that unites people, is analogous to it, and can enrich both parties in an analogous way (cf. Frasz 2001). Being a friend of the environment, according to Frasz, can benefit human beings more than exploiting it (Frasz 2001, 11). Hence, our previous attitude toward nature should be replaced by a friendly approach and the establishment of a relationship hitherto reserved only for humans.

The second strategy appeals to the benefits of the moral agent. A feature of virtues is that those who possess them usually derive certain benefits from them; in this case, this refers to the benefits we derive from the environment. The environment is not only a source of material resources but also provides aesthetic sensations and is a valuable place for physical, intellectual, and moral development. This very fact should promote the development of certain dispositions. Moreover, we should deepen those dispositions that will help us make the most of and enjoy the opportunities offered by the environment. Thus, by indicating which dispositions of the moral agent provide the opportunity to enjoy the various benefits of the environment, it is possible to determine which virtues are environmental virtues. Of course, this approach does not presuppose the unreflective exploitation of natural resources; rather, it refers to nonmaterial goods, those that the moral agent receives in contact with the environment.

The third strategy also relates to a certain kind of virtue, namely those that contribute to the human excellence of a moral agent. According to Sandler, environmental virtue is that which applies to human excellence. "On this approach, what establishes a particular character trait as constitutive of environmental virtue is that it makes its possessor a good human being" (Frasz 2001, 5). A fourth strategy for specifying environmental virtues and the desirable trait of an environmentally conscious person is to look at individuals who have played an exceptional role in protecting the environment. This is an appeal to the so-called paradigmatic character of a morally outstanding individual. Knowledge of the qualities, actions, and biographies of individuals involved in ecology can help identify environmental virtues that are desirable for environmental protection. Such a role as an exemplary figure or environmental hero can be played by someone who enjoys recognition in wide circles, or by a person of merit in the local environment. Observing these people and analyzing their deeds can help us assess which qualities are environmental virtues and will help the individual fully develop an ecological character. This strategy is not perfect as it has two limitations: "one limitation of the environmental exemplar approach arises from the privilege it places on obtaining beliefs about who is environmentally virtuous. To the extent that those beliefs can be distorted, narrow or otherwise inadequate, the approach can result in mistaken assessments of some character traits and an inaccurate account of what makes a character trait an environmental virtue" (Sandler 2006, 248). The second limitation is that this method "does not provide resources for adjudicating between competing beliefs about who is environmentally virtuous. The lives and characters of the heroes of North American environmentalists may differ substantially from those of the environmental heroes of North American sportsmen, ranchers, loggers or developers, as well as from those of people in other parts of the world" (Sandler 2006).

According to Sandler, these four strategies will help the moral agent understand what environmental virtues are and what virtues he needs to develop in a way that helps him effectively protect the environment. Although this philosopher gives four ways of determining environmental virtues, he also stresses that these ways are not mutually exclusive. On the contrary, many dispositions that are desirable for full development meet each of the given criteria, even though they are not perfect. According to Sandler, both the extension approach and the study of examples are useful tools in identifying environmental virtues and vices, but they do not avoid simplifications and errors (2005a, 3–6), some of which have been presented already.

5.2.3 The typology of virtues

Sandler's ethics is an individualistic ethics that emphasizes the lack of universal prescriptions and unambiguous codes, norms, and decalogues. In line with pluralist assumptions, it also avoids specifying virtue. This philosopher points out that not every virtuous person will act the same way in a given situation. This gives rise to his skepticism about moral exemplars. Sandler emphasizes the fact that an environmental role model is always culturally, historically, and socially conditioned; his behavior is appropriate in a given place, time, and context, but it is not necessarily worthy of emulation in another latitude and historical period. Such an approach makes the catalog of possible virtues very extensive. Despite these caveats, environmental virtues can be divided into six main groups (2007, 82).

The first group of environmental virtues is land virtues, inspired by Aldo Leopold's ethics, which make humans good citizens of the biotic community of life. According to Bill Shaw (1997, 53), Aldo Leopold's concept is an excellent basis for environmental ethics mainly because ecology as a science is concerned with the relationships between individual organisms. Hence, in ecology understood as environmental protection, we should look at relationships. A good starting point for analyzing the relationship between humans and the environment is to trace how nature has been addressed by a person who is not a philosopher but was involved in environmental protection. Such a person is Aldo Leopold, who was involved in environmental protection and whose thoughts have inspired generations of environmentalists, sparking a discussion about Earth ethics and inspiring a conversation about virtues for fostering ethical reference to the Earth. Examples of these virtues include love, considerateness, attunement, ecological sensitivity, and gratitude. Sandler alludes to the idea of expanding the scope of human moral concern and points out that when talking about man as a being pursuing his essential goals in society, it is important to keep in mind his connection to the wider community – man as part of the ecosystem (1997).

Seeing humans as part of a biotic community is one of the assumptions of Aldo Leopold, who writes in his book *A Sand County Almanac and Sketches Here and There* (1949) about the connections between all living organisms. This involves a change in thinking about the role of the human on Earth, who is dethroned and relegated to the position of a mere inhabitant of Earth, which he now has to take care of. This change in position is supposed to foster the emergence in human beings of feelings of love and care for Earth and should stimulate virtues that will encourage them to protect the planet. At the same time, it should eliminate those dispositions that cause an unreflective approach to the consumption of natural resources.

The second group is the virtues of sustainability, which promote the unity of the ecosystem so that it can produce the goods necessary for full human personal development. This group of virtues draws on the ideas of van Wensveen, who held that virtue is that which serves the sustainability of ecosystems. Sandler partly accepts van Wensveen's argument, emphasizing that "a disposition to maintain, or a disposition that tends to maintain, reliable availability of basic goods over the course of one's lifetime is justified, whereas dispositions that undermine their availability are unjustified" (2007, 44). However, he disagrees with the assumption that ensuring the sustainability of ecosystems is a prerequisite for virtue. He claims that ecosystem services[4] are important for ensuring the sustainability of the species and are important for the health of the moral agent, but ecosystem sustainability is not a prerequisite for the moral development of the individual. Even when an ecosystem is disturbed, destroyed, or degraded, this should not affect the moral development of the individual. What's more, it may happen that an individual obtains some goods from such an ecosystem artificially, which doesn't mean that he can't act in a virtuous way in the process. A disturbance or threat to the sustainability of the ecosystem should not be used as an excuse. Does a moral agent become incapable of acting properly under such conditions? Does his ability to act in a virtuous manner disappear and he immediately leaves the path of virtue? According to Sandler, an appeal to the sustainability of ecosystems is possible, but not in the version proposed by van Wensveen. This is because ensuring sustainability is not a prerequisite for an individual's moral development. Admittedly, Sandler does not assume, following van Wensveen, that ensuring ecological sustainability is a necessary condition for virtue, but he does assume that it is an essential element of it. Hence, dispositions that direct us toward consumerism are doubly harmful (2007, 56). First, a person with such dispositions will not achieve happiness, that is, he will not achieve the goal of eudaimonistic ethics. Secondly, indulging in consumerism leads to overconsumption of natural resources and thus undermines the sustainability of ecosystems. Sandler cites psychological research and shows that having more material goods is not associated with a belief in a high quality of life. Rather, greater satisfaction with life can be observed in people for whom social values and those related to self-realization are more important. According to him, people focused on material goals are characterized by a lower sense of well-being and poorer mental health. Although there is a correlation between a focus on material values and a subjective sense of well-being, it is difficult to explain the exact nature of this relationship. It is often stressed that setting materialistic goals and attributing significant importance to them involves compensating for deficiencies in other spheres, hence one should not pursue the simplistic statement that materialistic pursuits are a source of unhappiness and low quality of life. The issue is much more complex, and learning more about the nature of this correlation requires further research.

Examples of virtues in this group include temperance, frugality, farsightedness, attunement, or humility. Helpfulness is crucial in the case of these virtues and those associated with environmental activism. Being helpful on the ecological plane could guarantee preserving environmental assets for as long as possible and contribute to the judicious use of natural resources. Helpfulness is important in both its active and passive (not causing harm) forms (2007, 53).

The third group of virtues refers to moral aptitudes related to feeling unity with nature as these are dispositions that help a person enjoy nature and benefit from the environment. These include wonder, openness, aesthetic sensitivity, attentiveness, or love. These virtues are, in a sense, the calling cards of so-called environmental heroes who generally spend their time in nature and are focused on admiring it and trying to understand it. Their lives of connection with the environment are accompanied by a constant admiration for the wonders it provides.

Sandler, citing Rachel Carson, says that awe is the key to discovering the world, while at the same time it weakens the temptation to negatively impact the environment (2007, 50–51).

It is an environmental virtue in the full meaning of the word because it is an environmentally responsive, environmentally justifiable, and environmentally productive virtue. Wonder is a virtue because it leads us to right actions and supports the realization of good for the individual as well as the community, while also promoting the value of natural entities (2007, 51).

The natural environment enriches man and offers him a space to feel unity with nature, while providing a kind of atmosphere of renewal. However, man must be prepared to feel this bond with nature, or at least have within him a readiness for this kind of experience. This requires transcending the boundaries of selfishness and opening to the good beyond the personal good. Nature offers goods that are in some way unique, but the moral agent must be willing to accept them. This characteristic leads to a feeling of oneness with the natural world and is the source of the virtues associated with this experience.

The fourth group consists of virtues of respect for nature, including care, compassion, restitutive justice, nonmaleficence, and ecological sensitivity. The idea of respect for nature is well known to environmental ethicists as it is the cornerstone of many concepts and determines how to relate to the environment. In Sandler's case, it is very consistent with his concept of concern for the realization of an individual's goals as a representative of the species. Sandler emphasizes that humans represent a species distinct from all others, but he immediately adds that there is nothing extraordinary about our uniqueness (2007, 13). Every species is unique, has its own goals, and deserves respect and recognition. Sandler was very sensitive on this point. In his *book Character and Environment. A Virtue-Oriented Approach to Environmental Ethics* (64), as well as in later publications, he emphasizes the role of conservation efforts directed toward protecting biodiversity. Five years after the publication of that book, this philosopher published *The Ethics of Species* (2012), which is devoted to issues of species conservation with reference to the naturalistic approach presented in this book.

The best-known proponent of the idea of respect toward nature was Paul Taylor, to whose thoughts Sandler refers. The very fact that we are dealing with an animate entity makes it imperative that we give it respect as all organisms are teleological centers of life and as such deserve respect. According to Taylor, every organism pursues its proper goals and strives for what will allow it to best realize the potential inherent in it as a result of belonging to a particular species.

Moreover, every being has inherent and intrinsic value and is valuable in itself, but not because of its utility. Sandler recalls that, for Taylor, respect for life in all its forms is the foundation of virtue theory. It is respect for living organisms that creates a virtuous character. Moreover, Taylor considers respect for nature to be the most fundamental kind of moral commitment that can characterize a moral agent (Taylor 1986, 90). This requires making assumptions that are typical of biocentric ethics, that is, recognizing the value of life in whatever form it exists in the natural world. Sandler emphasizes that, according to Taylor, any rational, knowledgeable and enlightened person will adopt a biocentric worldview (2007, 66). This judgment lacks philosophical justification and is also somewhat fundamentalist.

Moreover, Sandler emphasizes that every animate entity is valuable as an object of our actions, and not all entities are moral agents; however, they all deserve our respect and consideration in moral choices (2007, 67). Moral agents are individuals who have the capacity for virtuous or non-virtuous actions, or they have duties and obligations. Nonhuman living organisms have neither such capacities nor such obligations. So, we should take these organisms into account in our moral choices, but they cannot be treated as moral agents.[5] Making arguments for the inherent value of all living organisms,

Sandler invokes the principle of the equality of species, as was formulated by Paul Taylor (Taylor 1986, 155). This principle assumes that regardless of what species a particular individual belongs to, it always has intrinsic value and deserves to be considered in moral choices by a moral agent. It is good in itself and is entitled to protection because of its intrinsic value.

The fifth group refers to the virtues of ecological activism, which are char-acter traits that support ecological sustainability. Among them, Sandler men-tions cooperativeness, compassion, commitment, optimism, and creativity. This is an interesting group of virtues related to human functioning in a social group (2007, 49). Although in his introduction to the concept of environmen-tal character Sandler accuses environmental ethics of having become a set of precepts and prohibitions similar to those present in law and politics, he appreciates the fact that involvement in society can influence political deci-sions. Moreover, environmental character can influence legislation, stimulate the formation of cultural patterns and the replacement of nonenvironmental ways of behavior with ecological ones, change the way the environment is perceived in society, or influence the direction of policy so that it promotes pro-environmental actions.

Virtues associated with environmental activism have been overlooked in previous discussions of environmental ethics, or at least they are not as popu-lar as the other environmental virtues discussed. Sandler blames the American way of thinking about environmentalism for the omissions in this group of virtues (2007). The established pattern of heroes as characters who wander in the mountains or through the woods defines the nature of environmental vir-tues, which foster an appreciation of the qualities of humans and the environ-ment. In this connection, the virtues of unity with nature, such as openness, humility, attentiveness, care, love, and so on, come to the forefront. This is an idyllic depiction of characters who spend time in nature, but it does not show the totality of their commitment to environmental issues because each of these heroes had a clear intention to convey a message of environmental protection. The pro-environmental activity of the heroes described was related to the desire to get other people interested in environmental protection (e.g., Henry Thoreau, Aldo Leopold), to put some areas under protection (e.g., John Muir), or to draw attention to the fact that our actions are harmful to the environment (e.g., Rachel Carson). The virtues associated with a mere appreciation of nature and its qualities are not enough to accomplish the tasks these heroes set for themselves in the area of environmental protection. Being a passive observer is therefore not the solution. Commitment and active environmental actions play a key role. Each of these heroes fought a battle of sorts to ensure that the issues they portrayed were properly recognized. Suffice to mention Rachel Carson, for example, whose criticism of technology brought a wave of negative comments. Although today her contribution to environmental protection and spreading knowledge about the impact of chemicals on the environment is appreciated, in her time the use of DDT was very common, and her bold claims were met with criticism (cf. Smith 2001).

The last group is the virtues associated with environmental stewardship, which, following Jennifer Welchman (1999, 411), Sandler understands as dispositions that contribute to being a good caretaker of the environment. This category includes benevolence, loyalty, justice, honesty, and diligence (Sandler 2007, 82). These are particularly important virtues for those in environmental roles as they express the concern of someone who for a period of time has been given custody of something valuable. Just as a museum curator cares for works of art and never claims rights to them, so a custodian of the Earth should care for it and its resources.

5.2.4 Summary

Sandler's concept of environmental virtue ethics proposes an expanded catalog of environmental virtues. The traditional elements of Sandler's concept refer to the views of the classics of environmental ethics, mainly Aldo Leopold and Paul Taylor. The innovative elements of Sandler's EVE, on the other hand, speak of the virtues of environmental activism, which mark a new space for thinking about environmental activism. Alongside this there are constant references to environmentally virtuous people, such as Rachel Carson, Henry Thoreau, and John Muir. Sandler, too, took very seriously the charge that there are no criteria for defining what is an environmental virtue and distinguished four strategies: (a) the strategy of extending the traditional interpersonal virtues; (b) the strategy of the personal benefits of the moral agent, according to which a virtue is that which benefits the person endowed with it; (c) the eudaimonistic strategy, according to which a virtue is a quality that makes a person good; and (d) the strategy of determining what, by observing environmental heroes, is an environmental virtue.

Notes

1 Ronald Sandler is employed at Northeastern University, where he is Chair of the Ethics Institute and is a member of research teams working on nanotechnology, society, and environmental justice issues. His main research interests include environmental ethics, ethics and technology, ethical theory, and Spinoza's thought. At his *alma mater*, he is a recognized academic. In the 2004/2005 academic year, he received an award for his teaching. He is the author of the following book publications: *Environmental Ethics. Theory in Practice* (2018); *Designer Biology: The Ethics of Intensively Engineering Biological and Ecological Systems* (co-authored with John Basl; 2013); *Ethics and Emerging Technologies* (book editor; 2013); *The Ethics of Species* (2012); *Character and Environment: A Virtue-Oriented Approach to Environmental Ethics* (co-author Phaedra C. Pezullo; 2007); *Environmental Virtue Ethics* (co-editor with P. Cafaro; 2005).

2 A peculiarity of Sandler's thought is the very clear reference to the works of contemporary virtue ethicists. Nevertheless, one can also see the strong influence of at least two environmental ethicists, Paul Taylor and Aldo Leopold, in addition to an apparent reference to contemporary works in the field of environmental virtue ethics, mainly Louke van Wensveen, Philip Cafaro, Geoffrey Frasz, and Jennifer Welchman.

3 An interesting paper on the implications of this terminology for discussion of specific virtues can be found in: Haught, *Environmental Virtues and Environmental Justice*, "Environmental Ethics" 2011, vol. 33, no. 4, p. 211–224.
4 Ecosystem services are *goods and services provided by ecosystems to humans.*
5 This distinction alludes to the difference, discussed in the philosophical literature, between a moral agent and being the object of moral choices (moral patient) (Cf. Gunkel 2012).

References

Annas J., *The Morality of Happiness*, Oxford 1993.
Aristotle, *Nicomachean Ethics*, transl. W. D. Ross, 2007, https://socialsciences.mcmaster. ca/econ/ugcm/3ll3/aristotle/Ethics.pdf (Access: March 17, 2021).
Birch T. H., *Moral Considerability and Universal Consideration*, "Environmental Ethics" 1993, vol. 15, no. 4, p. 313–332.
Cafaro P., *Environmental Virtue Ethics. Special Issue: Introduction*, "Journal of Agricultural and Environmental Ethics" 2010, vol. 23, no. 1, p. 3–7.
Calder G., *R. L. Sandler, Character and Environment: A Virtue-Oriented Approach to Environmental Ethics Columbia University Press, New York, 2007*, "Ethical Theory and Moral Practice" 2010, vol. 13, no. 2, p. 233–234.
Foot P., *Natural Goodness*, Oxford 2001.
Frasz G., *What is Environmental Virtue Ethics That We Should Be Mindful of It?*, "Philosophy in Contemporary World" 2001, vol. 15, no. 3, p. 5–14.
Gunkel D. J., *The Machine Question. Critical Perspectives on AI, Robots, and Ethics*, Cambridge MA 2012.
Haught P., *Environmental Virtues and Environmental Justice*, "Environmental Ethics" 2011, vol. 33, no. 4, p. 211–224.
Hill T., *Ideals of Human Excellence and Preserving Natural Environment*, "Environmental Ethics" 1983, vol. 5, no. 3, p. 211–224.
Hurka T., *Virtue, Vice, and Value*, Oxford 2001.
Hursthouse R., *On Virtue Ethics*, Oxford 1999.
Jonas H., *Zasada odpowiedzialności*, transl. M. Klimowicz, Cracow 1996.
Jonas H., *Seventeenth Century and After: The Meaning of the Scientific and Technological Revolution*, in: idem, *Philosophical Essays: From Ancient Creed to Technological Man*, New York 2010, p. 46–82.
Leopold A., *A Sand County Almanac and Sketches Here and There*, New York 1949.
MacIntyre A., *After virtue. A Study in Moral Theory*, Notre Dame 2007.
Rolston III H., *Environmental Virtue Ethics: Half the Truth but Dangerous as a Whole*, in: *Environmental Virtue Ethics*, R. Sandler, P. Cafaro (ed.), Oxford 2005, p. 61–78.
Sandler R., *Introduction: Environmental Virtue Ethics*, in: *Environmental Virtue Ethics*, R. Sandler, P. Cafaro (ed.), Oxford 2005a, p. 1–12.
Sandler R., *What Makes a Character Trait a Virtue?*, "The Journal of Value Inquiry" 2005b, vol. 39, no. 3–4, p. 383–397.
Sandler R., *A Theory of Environmental Virtue*, "Environmental Ethics", 2006, vol. 28, no. 3, p. 247–264.
Sandler R., *Character and Environment. A Virtue-Oriented Approach to Environmental Ethics*, New York 2007.
Sandler R., *The Ethics of Species*, New York 2012.
Sandler R., *Environmental Virtue Ethics*, in: *International Encyclopedia of Ethics*, H. Lafollette (ed.), 2013, DOI: 10.1002/9781444367072.wbiee090 (Access: November 29, 2023.
Sandler R., *Introduction*, in: *Ethics and Emerging Technologies*, R. Sandler (ed.), New York 2014, p. 1–23.

Sandler R., *Environmental Ethics. Theory in Practice*, New York 2018.

Sandler R., *An Interview with Ronald Sandler*, https://cup.columbia.edu/author-interviews/sandler-character-environment (Access: July 31, 2023).

Shaw B., *A Virtue Ethics Approach to Aldo Leopold's Land Ethics*, "Environmental Ethics" 1997, vol. 19, no. 1, p. 53–67.

Smith M. B., *"Silence, Miss Carson!" Science, Gender, and the reception of "Silent Spring"*, "Feminist Studies" 2001, vol. 27, no. 3, p. 733–752.

Taylor P., *Respect for Nature: A Theory of Environmental Ethics*, Princeton 1986.

Welchman J., *The Virtues of Stewardship*, "Environmental Ethics" 1999, vol. 28, no. 3, p. 411–423.

6 Narrative environmental virtue ethics

The author of the third concept of EVE is Brian Treanor,[1] who looks at environmental problems from a narrative perspective, referring to many philosophical and literary themes drawn from different cultures. This makes reading his book on environmental virtue ethics an interesting intellectual adventure. He claims that EVE cannot be just another ethical theory but should lead to the development of virtues that influence political decision-making (cf. Powell 2020). Besides, Treanor is also convinced that – as the nation that conquered North America – Americans have the most to do here. The colonization of America has left an indelible mark on the nature there and influences the thinking of contemporary US citizens, so it is necessary to take measures to not only address the problems in the natural world, but also to renew the spiritual connection with it.

The main inspiration and canvass for Treanor's views is Aristotle's conception of virtue ethics, which forms the core of narrative EVE and determines how virtues and the various elements of virtue theory are understood. Macintyre's thought is also an important inspiration in terms of how to understand the role of virtues. Treanor adopts the British philosopher's perspective of virtue ethics as an antidote to relativism, and to some extent he shares his communitarian views, as can be seen quite clearly in the context of the discussion of the social dimension of virtue. The reference to the concept of narrative may come as a surprise since narrative philosophy does not fall within the framework of virtue ethics. Nevertheless, in Treanor's account the concept of narrative is presented mainly as a tool for formal and informal moral education and for analyzing one's own moral deeds. Treanor's understanding of narrative and its place in ethics follows Paul Ricoeur.

6.1 The foundation of narrative environmental virtue ethics

The essence of the narrative[2] concept of environmental virtue ethics is the use of narrative as a tool for talking about virtues. A narrative can subtly tell the story of positive personal role models; thus it can prove to be an effective tool for encouraging moral agents to cultivate environmental virtues. As Treanor shows, it is crucial to select the right texts so that the full potential of narrative can be realized.

DOI: 10.4324/9781003433156-9

6.1.1 The narrative concept of environmental virtue ethics

Treanor accepts that environmental ethics cannot be complete without virtue ethics (2014, 22), and narrative is crucial in nurturing virtues, passing them from one generation to the next and creating a culture that incorporates environmental virtues. According to Treanor (2014, 109), narrative plays a huge role in understanding and expressing the human *telos* (aim) and contributes to its realization. This is because the fundamental question that man asks himself is about who he is. Following the example of the ancient *gnoti se auton*, modern man is constantly trying to understand himself and discover his own identity. This search in environmental ethics is accompanied by questions about what kind of man an individual should be and wants to become. Considerations of this kind are an essential element of narrative, or, in the words of Paul Ricoeur, narrative identity. Moreover, changes flowing from ethical motives are accompanied by a change in narrative. As Treanor writes, "when I change the person I am – as when I attempt to become virtuous – I'm 'refiguring' the narrative I am" (2014, 114).

Why the narrative approach? Because, according to Treanor, narrative plays a key role in ethical formation, whether it is the formation of the characters of children or adults (2014, 22). As Treanor writes, "good people can make good use of rules. Rules, guidelines, adages, proverbs, and aphorisms (…) provide a useful ethical 'shorthand' for expressing what a good person would generally do in most situations" (2014, 157). Narratives tell us what is right and wrong and support the use of prudence in making one's own choices. The plot of a narrative is a mythos – a sequence of events that make up a story – but at the same time the plot is the place on Earth or the environment in which the story takes place (2014, 22).

At the same time, because it uses thinking with images, a narrative can be more interesting and more easily remembered than a scientific lecture or a sermon. The tools of narrative are symbols, metaphors, or images conveyed through myth. Myth plays a significant role in culture; it is part of archaic cultures, but Plato brings it back to philosophy because of its great educational potential and its ability to convey complex phenomena through images and metaphors. The most famous philosophical myths can be found precisely in Plato, where they have several functions: myth-comparison; image-example (e.g., in the form of a historical legend) and allegorical parable (e.g., in the form of a historical legend) (cf. Wolicka 1994, 58). According to Wolicka, "myth in Plato's philosophy is not a form of cultural relic – a historical relic inherited from the archaic tradition – but a consciously introduced and controlled method of expressing and interpreting metaphysical truths" (Wolicka 1994).

Moreover, metanarratives, such as religious narratives, precisely use metaphor, image, and myth. For example, the aforementioned story of the battle between the Pandavas and the Kauravas is nothing less than an epic about the struggle between good and evil inside each of us, while the parable of the merciful Samaritan is a message of Christian love toward our neighbor. Treanor cites several novels and myths that are treasures of the cultural heritage of many traditions and at the same time teach wisdom and portray virtues. Most importantly, these very representations operate with images that can be easily remembered, and they utilize mediums that allow one to readily absorb the lesson implicit in a given narrative.

Treanor draws his inspiration for the creation of virtue ethics mainly from Aristotle, whose views in this area he considers very mature and still relevant, despite the passage of time. However, he stresses that he is not faithful exactly to Aristotle's ethics, but rather to its spirit (2014, 25–26). The essence of his ideas is the answer to the question of what it means to be a good person. Treanor's ethics is a teleological and eudaimonistic ethics whose task is to help the moral agent achieve the most important goal of life, which is lasting happiness, made possible by virtue and life fulfilment.

What then is true happiness? Treanor, following Aristotle, argues that the substructure of lasting happiness is reasoning. Reasoning and reasonableness are among the key elements of virtue theory. William Prior (2001) emphasizes the central role of reasoning in Aristotle's virtue ethics, while pointing out that for many virtuous acts the moral agent relies on the ability to reason. An act can be virtuous due to the use of the intellect in analyzing a particular situation. The use of reason itself contributes to the development of the moral agent and is therefore an element that develops a person. Consequently, the usage of reason fosters an individual's pursuit of full personal development, which is one of the elements of eudaimonistic ethics. Indeed, the Greek *arete* is a broad concept; the pursuit of a truly happy life means striving for perfection and fostering human development at the level of morality, intellect, and corporeality (Treanor 2014, 28). Human well-being is not narrowed only to the moral sphere; Treanor here adopts from Sandler a naturalistic understanding of the good.

6.1.2 *The importance of narrative*

Narrative is the main tool for moving from who I am to who I can become. It is an instrument to help one become the best version of oneself. Used skillfully, narrative helps to overcome the preachy tone of many discussions about virtue. It allows one to talk about virtue lightly and yet more effectively than in the manner of moralizing treatises on living a virtuous life. A well-composed narrative has genuine power to influence the conscience of the moral agent and thus affect his fate.

In the first place, narrative helps to answer the Socratic question: who am I? It helps to discover narrative identity (Ricoeur 1992, 110), facilitating discernment in one's own moral prowess. Virtue ethics forces one to ask questions about one's interior: what kind of person am I? What dispositions do I have? It is this process of self-reflection through inner narrative that is the source of inspiration for cultivating virtues, for personal improvement or transformation. The process of self-reflection and telling oneself the events of one's life helps to create a story, writing – in a metaphorical sense – the narrative of one's life. It is a transition from the question "Who am I?" to the question of "Who will I be or who do I want to become?" Ricoeur wrote about the formula for transforming one's life as a process consisting of the following stages: description, narration, prescription. This last element makes it possible to decide not only on the actions taken, but, as it were, it constitutes the moral agent – it creates oneself.

Narrative allows us to create ourselves and write our own story. In a metaphorical sense, since we do not have full control over the course of our own lives, we are actors in it who face what they encounter without prior knowledge of the script of events. Nevertheless, we are both protagonists and narrators of our own lives, and in some limited sense we can even make conscious decisions that change the direction of our lives. At the same time, being the narrator of one's own life requires us to be honest and able to realistically assess our own capabilities. The image of ourselves must not be clouded by the illusory notions in which we often indulge. Narrative should purge us of them, for it organizes our thinking and gives it meaning. As Hardy notes: "we dream in narrative, daydream in narrative, remember, anticipate, hope, despair, doubt, plan, revise, criticize, construct, gossip, learn, hate and love by narrative" (1968, 5). In narrative, we are able to know ourselves better and read the events of our lives, seeing in them a certain continuity between the past and the present. Narrative also helps us realize teleological ethics by facilitating the perception of the purpose of moral life.

A key element of the narrative concept of virtues is the motivation to become the best version of oneself. As Treanor notes (2014, 161), each of us has some image of ourselves that we strive to realize, and this image is shaped by narrative, which influences how we understand goodness and the good life. This image of a moral agent's high quality of life is shaped by literature, proverbs, myths, as well as the mass media. As an ideal example of thinking about narrative, personal transformation, and virtue, Treanor cites the character of Don Quixote, the hero of Cervantes's novel. At the same time, he proposes rejecting traditional, templated ways of interpretation and, instead, looking at this character from a perspective different from that perpetuated in the culture. Don Quixote is a perfect example of succumbing to patterns from narrative; his madness and lack of reason stem from his fascination with chivalry novels. It is literature that makes the novel's protagonist want to become a misguided knight, that is, a warrior who wanders in search of adventure and defends the lady of his heart. His inability to rationally assess the situation makes Don Quixote a tragicomic figure. Treanor points out that in Don Quixote's case this irrational assessment of the situation, driven by dreams of becoming a knight, is due to his lack of knowledge of literature other than chivalry novels. It was literature that caused Don Quixote to run the narrative of his life along the lines of the books he knew, rather than taking his own circumstances into account.

As an example of another character who is led astray by literature, Treanor cites Christopher Johnson McCandless, the protagonist of the best-selling novel *Into the Wild* (Krakauer 1997), which was adapted into a movie.

The book, based on fact, describes the life of an American hiker who abandoned civilization and hitchhiked to Alaska, where he survived for more than 100 days in the forest wilderness; however, after 113 days he was unfortunately found dead in an abandoned bus. It is assumed that McCandless's expedition was largely inspired by the works of Henry Thoreau, Ralph Emerson, Leo Tolstoy, or Jack London, who called for spiritual search, uniting with nature, and rejecting social norms. McCandless, the protagonist of *Into the Wild*, wanted to live by higher ethical standards, so he began his journey by donating his own college fund to the charity Oxfam. He wanted to go beyond the usual consumerist model of life that most young Americans adopt. Unfortunately, the ideals he believed in fell short of his potential, and he became a victim of his own notions of an ideal life.

Even though characters such as McCandless face criticism, they embody virtues. Treanor stresses that the understanding of the virtues found in Don Quixote must not be distorted by emphasizing their weaknesses and flaws. It is important to note what is valuable: for example, in the character of the Lord of the Mancha, it was his genuine devotion to love and justice. McCandless, on the other hand, embodies the quest for self-discovery and exemplifies a deep commitment to truth and a rejection of the consumerist model of life.

In addition to these narratives, which according to Treanor carry a positive message anyway, there are also narratives that have an even clearer message, such as the aforementioned *Walden; or, life in the woods*; however, Treanor also refers to the works of other well-known American authors, such as John Muir (1901, 2005), David Brower (2000), Edward Abbey (1985), or Doug Peacock (1996), whose books are literary descriptions of the beauty of nature, mainly areas largely or totally unaffected by human activity. All these authors praise the bond with nature and express respect for it, and above all their writing could serve as a narrative that evokes love for nature. These authors could trigger a change of attitude in readers and make them want to change their lives through reading. This was the case with Rosalind Hursthouse, who, influenced by reading philosophical texts, decided to switch to a vegetarian diet (Treanor 2014, 164). This is one example where a narrative translated into a lasting change in the life attitude of a moral agent.

Research shows that reading develops empathy. Narrative can play a role analogous to priming, "the effect in which recent experience of a stimulus facilitates or inhibits later processing of the same or a similar stimulus" (APA dictionary). Priming is a significant element in shaping the action of a moral agent and is an interesting mechanism used in social psychology. This mechanism is familiar to salesmen and politicians, who use it to either increase sales or gain favor with voters. An appropriately chosen narrative could play an analogous role and become a catalyst for desirable moral attitudes.

To date, priming has been mainly associated with the situationist critique of virtue ethics, which attempted to show that virtues are not fixed dispositions. John Bargh (1996; cf. Szutta 2015, 29) attempted to prove that situational factors of various kinds have a significant impact on our behavior that we generally do not fully realize. This type of research, according to situationists, undermines the thesis of virtue as a fixed disposition or character trait. Virtue ethicists do not agree with the claim that situational factors have a significant influence on our behavior (cf. Szutta 2015; Szutta 2012; Jaśtal 2015). Nevertheless, narrative can guide the behavior of a moral agent in a manner analogous to the process of priming[3] (APA 2018). In Treanor's view, however, it is not priming in its pure form because its effect on the moral agent should be relatively constant, and the choice of texts should be conscious. Thus, only in a certain sense is narrative similar to the process of priming.

Narrative in this view has an impact not only on individual development but also on society. Every culture has a series of stories that use the power of imagination to teach about what is right and what is wrong. In our culture, examples include Aesop's stories, while in Far Eastern cultures it might be the Buddhist Jataka tales[4] or the Hindu Panchatantra.[5] Each depicts virtues that are particularly valued in a given culture and colors these in a way that is specific to the cultural background of the narrative. The strength of some narratives lies in their reliance on paradigmatic moral characters, as is particularly evident in the stories of the life of the Buddha, whose life has inspired generations of Buddhists. Each culture, therefore, has its own stories and myths that can play a role in the transmission of virtues. This diversity of role models and attitudes is a great value of narrative and is not an obstacle to emphasizing the universalist nature of virtues.

Treanor (2014, 166) refers to the words of Plato, who argued in *The Republic* that narratives are used to educate children because a child's reasoning ability is not sufficiently developed. Treanor writes, for example, about a story depicting a boy who warned a village about a wolf as a joke. Repeated several times, the 'joke' ensured that when a real threat arose, no one took the boy's warning seriously. Critics of the role of narrative emphasize that it can help make the moral agent aware of how to act, but it says nothing about the reasons for doing so. The motivation for telling the truth is not fear of wolves, but the belief that lying is wrong as an act and has worse consequences than telling the truth. However, Treanor notes that this type of argumentation stems from an underestimation of narrative and an overestimation of the self-sufficiency of reason. In his view, narrative plays an important role in the ethical formation of both children and adults and is a valuable complement to legalistic modern ethical concepts that are focused on formulating norms.

Treanor pays particular attention to the role of narrative in environmental issues. On the one hand, the response to climate change is limited because the human brain is designed in such a way that humans react to an immediate threat rather than to phenomena that develop over a long period of time, and climate change is one such phenomenon. The change in living conditions on Earth resulting from climate turbulence is occurring at such a pace that it does not yet cause an immediate threat to human life (cf. Goleman 2009, 37). However, the occurrence of extreme weather events and anomalies has increased over the past few years, and the shifting of seasons or other negative effects of climate change are also evident. Thus, an appropriate narrative, guided through education, can help integrate the climate sphere and other Earth ecosystems into our moral reflection.

Narrative can be used to rediscover nature; as Treanor argues, contact with pristine nature shapes us as moral agents and builds our character (2014, 3). It also influences the nation we live in, creating its history, which means that when we lose wild nature, we become different people. Such is the fundamental importance Treanor attributed (after Wallace Stegner)[6] to pristine nature in human life. At the same time, this relationship is reflected in language and literature and creates specific narratives that shape culture as "our language and literature reflect a constellation or network of relationships where individual identity, social identity, the world and the environment intersect and are entangled" (2014, 4). Both literature and other forms of artistic expression reflect the state of humanity and its relationship with the natural environment.

According to Treanor, narrative can serve to shape our attitudes in the future. This philosopher cites Mary Gergen's (1995) research showing how the literature of a particular period influences the character of romantic relationships. Literature, myths, and novels can determine attitudes in relationships between people and the way we view intimate relationships. Our narratives about ourselves influence our understanding of how we perceive the reality around us and create images that form our perception.

Nowadays, many narratives come from images provided by mass media. According to Treanor, mass media should aim to convey environmental virtues as images are more effective in shaping attitudes than dry facts, information, or data. The right narrative can help show the moral agent how to cultivate environmental virtues. Treanor points out that in every era there is a need to nurture certain virtues that are in deficit: for example, the 21st century is a time when it is necessary to nurture environmental virtues (2014, 186). While preserving other essential virtues, we should therefore focus on nurturing the moral prowess that will help us face the environmental crisis.

6.1.3 Phronesis – practical wisdom

Virtue ethics does not offer ready-made prescriptions on the question of how a person should take appropriate action (2014, 87). Moreover, we have known since Freud that the self is not completely transparent to itself. Therefore, even we ourselves may find it difficult to recognize whether our action was indeed virtuous (2014, 88). In addition, examples of the application of virtue vary from person to person: what is an act of courage for one person may not be so for another. This imprecision of virtues and their individual character make it possible to fall into the trap of relativism. Aristotle's answer to this challenge is phronesis, namely practical wisdom which is supposed to help the moral agent make the right decisions, especially in ethically complex situations. The way practical wisdom is applied in modern times should be different from what we know from ancient philosophy. Relativism, which has always seemed tempting, is now reinforced by the phenomenon of globalization, which weakens the importance of grand narratives.

As is the case with Aristotle and modern philosophers who deal with virtue ethics, a key role in Treanor's work is played by *phronimos*, through which the moral agent has the ability to take the right action in the face of a conflict of values. *Phronimos* is the guarantor of right decision-making and is the source of the formulation of norms that is characteristic of virtue ethics (cf. Dzwonkowska 2017). The key role in the pursuit of realizing the good and being virtuous, according to Treanor, is played by two skills: discernment (the ability to recognize vice and virtue, and the ability to recognize the context in which they occur) and understanding (insight into the nature of realizing one's potential, virtues and vices) (2014, 169). According to Treanor, narrative is fundamental here, for literature tells the moral agent to "change your life!" but at the same time provides several possible directions for such change. Changing one's life needs the ability to make good choices: "discernment and wisdom are both the result of experience. Aristotle says that those who study ethics and politics should have experience with life, and he later says that practical wisdom (*phronesis*) is related to having experience in the world" (2014, 170).

Narrative helps to forge and refine discernment, which, like wisdom and prudence, is based on an 'as-if' experience that helps us to experience a situation in a certain way. This ethical experiment gives us the opportunity to understand events in their various aspects. Ricoeur describes narrative as a kind of imaginative or ethical laboratory in which we can analyze various situations and actions. The knowledge that flows from this laboratory is a kind of peculiar experience that helps us build knowledge about ethical action. In this way, the 'as-if' experience helps us build knowledge about ethical action and provides us with the information we need to make decisions that lead to good actions; at the same time, it helps us to resist modern relativism.

6.1.4 Narrative as a response to relativism

Phronesis is crucial to overcoming relativism, and its auxiliary tool is narrative. Treanor notes that the globalization and homogenization of culture is accompanied by the disappearance of metanarratives that are linked to membership in a particular religion or nationality (2014, 92). The loss of relevance of metanarrative, and at the same time the inability to create a universal metanarrative that would fit all cultures, is a challenge for virtue ethics. Treanor cites John Caputo (2000), who emphasizes that Aristotle's virtue ethics was intended for a closed, homogeneous, aristocratic society in which every virtuous person would have the same set of virtues as Aristotle. The metanarrative of homogeneous Greek society determined these virtues. In the current situation, where we are skeptical of metanarrative, it is difficult to even say what it means to be virtuous. Metanarrative determines *telos*; thus, the absence of metanarrative leads to the absence of the goal of ethics and the impossibility of developing an ethics of virtues (Treanor 2014, 93). The postmodern distrust of narrative undermines the possibility of developing virtue ethics adequate to our times, but it is not an impossible task. In his paper, Treanor claims that

> two sorts of environmental narratives, working in concert, further help to limit relativist objections: (1) narratives of environmental survival (which identify dispositions, such as simplicity, which are necessary for our long-term survival), and (2) narratives of environmental flourishing (which make a virtue of necessity by pointing out that those dispositions necessary for our survival often contribute to our flourishing beyond mere survival).
>
> (2008)

However, later in his book, Treanor mentions contemporary philosophers who offer responses to the threat from nihilism and relativism, namely Martha Nussbaum, Ronald Sandler, and Alasdair MacIntyre.

Martha Nussbaum proposes a relativistic reading of Aristotle. Her starting point is to note that many thinkers have trouble identifying a single universal principle of eudaimonism that could be applied by all. Most such philosophers refer to the possibility of multiple local norms. Nussbaum points out that Aristotle, too, never proposed a single universal indicator of human goodness; rather, he indicated spheres of action in which each person must make choices in a given area, so it is worth considering what constitutes a good or bad choice. Even when our answers on this issue differ, they still concern the same sphere of functioning of virtues, so they cannot be so far apart as to form incompatible discourses (2008, 100).

Nussbaum proposes an approach based on consideration of capabilities. This approach, developed in collaboration with Amartya Sen, makes it possible to focus on what people can do and who they can be, thus ensuring their dignity (Nussbaum 2001, 5). Nussbaum and Sen are of opinion that their concept is superior to utilitarian-economic and cultural-relativist approaches, which offer inadequate solutions to women's problems in developing countries (Jayawardena 2001, V). Their concept is of great practical importance because it allows women to use their potential to achieve real change in their social situation. Nussbaum and Sen's concept is based on moving away from the question of women's quality of life in comparison with that of men and replacing this question with another: what can the less privileged do and what can they become? It also moves away from including GDP in assessing quality of life, since the correlation between this indicator and society's quality of life is questionable. This approach creates the opportunity to realize one's own potential and consequently gives individuals the chance to become independent of aid programs (except for the initial phase, when individuals need support with using their abilities to be able to earn their own living).

The ten main capabilities include life, health, bodily integrity, practical reason, belonging, and the right to entertainment and control over one's political and material environment (Treanor 2014, 101). However, critics note that the criteria of capabilities and the concept of the good life on which Nussbaum bases her reflection are built on the notion of the good life as understood in the Western world (Cielemęcka 2011, 185). The imposition of the Western way of life is evidenced, for example, by the attribution of key importance to the issue of respecting women's rights. In developing countries, on the other hand, women often face problems that are more existentially burdensome than, for example, lack of education. This argument needs consideration. However, according to Treanor, Nussbaum's approach allows us to define the basis of a concept that gives a view of what is important but at the same time is not a rigid dogma that is detached from life. It is a theoretical position that is built on the Aristotelian model, that is, it offers universal truths with which to consider a specific situation. In this way it avoids the Scylla of absolutism and the Charybdis of relativism (Treanor 2014, 102).

The second approach that Treanor believes does a good job of answering the question of modern relativism is Ronald Sandler's environmental virtue ethics, or, more specifically, his naturalistic, pluralistic, and teleological approach to the issue. The concept itself is discussed more extensively in the fifth chapter of this book, so here I will only focus on its features that are necessary to overcome the relativism of modern culture. First and foremost, the key here is a pluralistic approach that broadens the circle of entities considered in moral choices and does not narrow the concept of eudaimonism to concern for one's own well-being. Moreover, going beyond concern for one's own well-being does not mean focusing only on personal entities, but also concern for the entire biotic community.

Sandler stresses that the very practice of environmental virtues contributes to the development and personal excellence of the moral agent, and it is also of importance to the environment. The definition of environmental virtue is that it is a character trait that benefits the moral agent and their environment (understood as the biotic community). The naturalistic view of virtue ethics emphasizes the biological nature of the moral agent and makes visible his relationship with the natural environment. At the same time, it does not forget concern for the social dimension. Sandler's concept outlines a broad framework for thinking about humans' moral obligations, and its boundaries are broad enough to guard against relativism. At the same time, pluralism and the broad framing of the issue help avoid absolutism in ethics. According to Treanor, the norms derived from Sandler's pluralistic, naturalistic, and teleological EVE raise questions about the best possible conditions for the development of man and his natural environment.

Treanor proposes MacIntyre's concept of virtue as a third response to the challenges posed for virtue ethics by relativism. In *After Virtue: A Study in Moral Theory*, this philosopher criticizes emotivism and relativist ethics, accusing them of being mere expressions of the individual preferences of the moral agent. He stresses that the communitarian importance of virtue is determined by a community of persons and is developed in that community. However, the multiplicity of communities and their differences, according to

Treanor (2014, 104), leads to relativism. Moral relativism itself was born precisely from the perception of divergence between supporters and opponents of moral views and the impossibility of reconciling them. This problem concerns both ethical issues (the meanings of the terms "virtue," "vice," "moral goal of human life" – *telos*) and how to justify one's ethical position. As Treanor notes, at the core of this gap is a deep commitment to truth and rational justification of ethical theses. Making an attempt to rationally justify one's ethical views confirms that the moral agent accepts the following theses (2014, 105):

1 The moral views he defends are not tainted by his own situation, that is, the moral agent attempts to defend the facts as they are, not as they appear to be to him.
2 "That moral standpoints incompatible with the belief in question are somehow flawed, that is they can and should be replaced by a rationally superior standpoint" (2014).
3 If in further inquiry, discussion, or experience the moral agent discovers that he cannot better rationally defend his views than his opponent, he should replace the proclaimed view with one that can be justified rationally.

The desire for rational justification is driven by a love of truth and a desire to create an ethics that is adequate to the world surrounding the moral agent. Devotion to the truth also implies a willingness to change one's own views when another ethical position offers a more convincing justification. This is problematic for MacIntyre, who, on the one hand, emphasizes devotion to one's own community and confinement within its cultural circle; on the other hand, he requires the moral agent to transcend his own perception of reality and redefine his moral position when he discovers more-convincing views based on rationality. Therefore, relying on rationalism in formulating moral judgments requires adopting a certain conception of truth. In addition, one must then constantly and systematically subject this conception of truth to criticism that reveals its limitations. The last thing necessary in this case is to question the conceptual framework of relativism (McIntyre 1988, 166). This means being open to reformulating one's ethical views as a result of contact with others and their worldviews.

These three positions, according to Treanor, respond to the challenge posed by the relativism of postmodern culture. They provide prescriptions that are not relativistic but at the same time are open to recognizing the individual dimension of virtue. They recognize the diversity of moral agents and allow culture to free itself from relativism. Each of these concepts outlines a conceptual framework that delineates a space for making one's own moral decisions according to one's individual goals (as in Ronald Sandler's case), abilities (Nussbaum), or a conception of truth determined by the moral context in which a particular agent lives.

6.1.5 Moral education

According to Treanor, the role of narrative is best seen in moral education. The emphasis on practicing virtue has made moral education one of the pillars of the narrative concept of EVE. This is because narrative plays an important role in inspiring the moral agent, motivating the cultivation of specific virtues, and transmitting moral models linked to virtues. Thus, narrative becomes a tool for transmitting patterns, but also for shaping the individual's ability to distinguish between right and wrong and to make a moral assessment of a given situation. Narrative plays a huge role in motivating us to become the best possible version of ourselves. Treanor devoted considerable attention to this issue, giving examples of heroes from literature whose fate can inspire readers to change their own lives. Narrative here serves as a useful tool in the informal education of adults and young people. Promoting good role models can also be done through inspiration from sources other than books. What is important is the role narrative plays in shaping certain desirable cultural patterns. In addition, it is crucial to involve children in the pursuit of goodness and in the desire to become a better version of themselves and change their habits.

Through narrative, the process of educating children and adolescents and re-educating adults is possible. This very feature of narrative gains a special place in the thought of Treanor, who draws attention to the role of ethical education carried out through narrative and its role in inspiring personal *metanoia*, in motivating us to become better, and in transmitting the values inherent in our cultural circle (2014, 160). An aspect of narrative is the process of internal narration, which makes us consider various situations and think about what we would do and how the situation could develop. Such internal narration allows us to learn the truth about our own character and the virtues and vices we possess (2014, 189). In addition, it helps us reflect on and understand the meaning of our lives, thus individual events can be seen as part of a larger whole.

According to Treanor, narrative can play a key role in the personal transformation of the moral agent (the Greeks called such a transformation 'metanoia'): narrative can inspire *metanoia* and show the direction of personal development. A crucial issue is the selection of narratives appropriate to the challenges of the 21st century. As Treanor notes by citing a letter written by Stegner regarding the protection of pristine nature, "ailing, embittered, and faithless aspects of our literature are manifestations of a broader illness and malaise stemming from a distorted relationship with the environment" (2014, 4). Thus, the choice of appropriate narratives is crucial. The criterion for selection in this case is the possibility of evoking this *metanoia*. This is precisely Treanor's goal. Although, as he points out, *Emplotting Virtue: A Narrative Approach to Environmental Virtue* Ethics is an academic book, his goal – following the example of the ancient philosophers – is to help people lead better lives through virtue (2014, 23). In his view, the highly advanced specialization of modern philosophy has contributed to the fact that philosophy no longer poses general questions. Scholars analyze complex problems that are increasingly different from everyday life issues (Treanor 2017, 201). Meanwhile, it is equally important to ask questions about general issues, such as "How to be a good person?", the answer to which should be understandable to every educated person.

Narrative plays an important role in transmitting ethical values to children as it motivates them to be good whilst also showing what it means to be good. Treanor points out that every cultural tradition has its own stories that convey moral truths. In Western societies, Aesop's tales play an important role, while Far Eastern countries have the Buddhist Jataka stories and the Hindu Panchatantra. Each culture has its own set of narratives that convey the values of a particular tradition and show ways of realizing them. This demonstrates "the genuine ubiquity of moral instruction via folktales, fairy stories, myths, and other narratives" (Treanor 2014, 166).

Stories that on the surface appear to be entertainment for children nevertheless also contain messages that are quite adult. Treanor cites the words of John Ronald Reuel Tolkien because in his tales of enchanted lands – a 'perilous realm' inhabited by all manner of archetypal fairies, giants, elves or dragons – there is teaching on how to be good. In this way, unreal events become elements of moral patterns in real life. According to Tolkien, it is possible to influence morality because these stories are in some way allied with reality. Although they are fantasy novels, they reflect reality and influence it – as does narrative, which reflects life.

The great moral messages of the founders or major figures of religions took the form of stories. One need only mention the parables of the Bible or the *Bhagavad-Gita*. Both the Bible and the holy book of Hinduism use stories to convey the truths proclaimed by spiritual and religious leaders. The characters that appear in biblical parables and in the *Bhagavad-Gita* show how to act and what to avoid. Both these examples provide guidance to believers in a simple, pictorial way that undoubtedly appeals to them more than abstruse philosophical or theological treatises.

Narrative supports the learning of virtues, which takes place in a manner analogous to the acquisition of physical strength (2014, 29), that is, a given virtue must be developed and maintained, just as exercise develops and maintains bodily fitness. Virtue, according to Treanor, is a matter of habituation – upbringing. In the case of children, this habituation is not just education in specific virtues but also involves creating a desire to be good and virtuous because the individual "must be brought up in such a way that being good matters to her, for if she is not, it seems highly unlikely that she will concern herself with it as she continues to develop" (2014). Narrative can assist in the process of forming and developing virtues; it can be a tool to promote personal development, which serves not only one's own good but also that of society as a whole.

An interesting argument in favor of narrative is the emphasis on the role of fictional events in the formation and transmission of moral patterns. Some messages come from stories, myths, folk legends, and fiction, while the examples shown earlier come from the books of the major religions. The stories presented in them are not always descriptions of situations that happened, and even if they reflect actual events, they are sometimes embellished. It is this element that is important in narrative, as Treanor writes, "fictional events can be imbued with a power and pathos it is difficult to find in ordinary experience" (2014, 173). Pathos, therefore, fills the content of the message, becoming apparent in certain motivations and emotions that can help shape moral behavior. Treanor argues that these emotions and the colorfulness of experience are illustrated in narratives, while the narrative of our lives is much more colorful than the real experience itself that we gain on earth.

Narrative plays a key role in shaping our beliefs, and these beliefs in turn contribute to certain actions, which often develop habits (Greek *hexis*) in us. Virtue is a permanent disposition (*hexis*): it is not enough to do the right thing once to become a virtuous person. This is where the importance of narrative comes from: not only as a tool for retrospection or analysis of past events, but also as an instrument in the formation of virtue. Thus, narrative is a virtuous character-building element that shapes our identity but also determines the nature of our behavior and moral choices. "Consciously or unconsciously, we act out narrative roles" (2014, 182). Hence, in ethics, including environmental ethics, we should be aware of the importance of narrative in understanding reality and guiding human behavior. Narrative determines a person's worldview and influences what a moral agent thinks about the reality around him. The doubts that have grown up around climate change are a good example for showing how a moral agent's narrative influences the perception of scientific facts.

In Treanor's view, narrative is the foundation of ethical formation and contributes to the moral agent's acquisition of practical wisdom. "A person won't achieve virtue in the dusty stacks of a library or archive, but in the choices, affects, habits, and dispositions of her actual life" (2014, 175). Man does not become perfect by merely reading books.

6.1.6 The 'as-if' experience

The 'as-if' experience plays a crucial role in narrative. It is a quasi-experience that allows one to experience a moral situation by empathizing with the events encountered by the character of a narrative. In Treanor's view, contemporary ethics are too removed from everyday situations, and philosophers speak in abstruse language that is even incomprehensible to an educated person. Literature, however, provides something like an "ethical laboratory" (Ricoeur 1984, 59), for it triggers our imagination and allows us to live the experience 'as if.' This type of approach is not new in ethics; for example, Aristotle wrote of the role of poetics in showing the essential truth about man (cf. Kearney 2002, 94). As the Stagirite points out, poetry itself devotes a great deal of space to the human being in his or her most important life experiences; it conveys lessons about human nature, but it also allows us to empathize with the protagonist. It provides an opportunity for a thought experiment involving the question "What would I do in a given situation?" The plot of a text involves us in thinking about the situation of the literary character, but through this it has a transformative power. It puts the viewer in the position of analyzing the story being read and asking questions about the moral situation in question. Above all, the reader is transformed into a literary subject, so to speak, and reflects on the solution to the problem faced by the protagonist of the work being read.

The ethics laboratory allows us to exercise ourselves in all kinds of moral situations and broaden our experiences much more effectively than in everyday life. In this way, we deepen our understanding of ethics and gain experiences that we would never get in real life. Treanor evokes the words of Martha Nussbaum (1990, 17), who, in relation to practical wisdom, argues that we can never experience enough, so 'as-if' experiences ensure that we gain moral knowledge and broaden our horizons. According to Treanor, the moral agent can choose any narrative, thus experiencing a completely different way of life than his own.

The example of a virtuous person from whom the moral agent learns virtue has an analogous effect. When faced with a choice, the moral agent should ask himself "What would I do if I were an exemplary moral character?" "What would I do if I possessed the virtues that he or she possesses?" The question, then, is not what the ecological hero would do, but what I would do if I had the virtues he has. This constitutes a form of transferring the experience of 'as-if' to oneself and obtaining an answer tailored to the moral agent's capacities. A narrative teaches independent moral decision-making, which is not a simple copying or imitation of a moral ideal but leads to individualized moral solutions.

The task of a role model of virtue is thus seen in this case as analogous to that of Plato's "divine hermeneutic" (Wolicka 1994, 26). The crucial aspect is *mimesis,* or the concept of imitation, which governs the emergence of phenomena in the physical world, including works of art and literature. A proper reading of the objects that represent reality in myths requires a kind of preparation for "seeing the essence of being" (Wolicka 1994) and understanding the virtue in this. Ricoeur speaks here of three ways of understanding mimesis: "1. it is a pre-understanding of human action; (…) 2. a literary configuration, which transforms events into narrative (…) 3. the intersection of the world of the text with the world of the receiver, the product of interpretation" (Rosner 2002, 136; cf. Ricoeur 1984, 62).

The third understanding of mimesis has ethical potential, meaning the power to change a moral agent and make them begin to reflect on the meaning of what they have read. Hence,

> When narrative opens up an imaginative 'as-if' world or ethical laboratory, it not only allows us to imagine what it would be like to be *another person*, it also allows us to imagine what it would be like to be *another version of ourselves.*

(Treanor 2014, 177)

The formation of virtue requires creative imagination and the refinement of moral values provided by an appropriate narrative.

The 'as-if' experience is nurtured by our ability to perceive real and symbolic phenomena in a similar way. As Treanor writes (2014, 173; cf. Sapolsky 2010), the moral distaste evoked by the cases described in the literature is perceived on a neurological level in the same way as that experienced in everyday life. Our brain is constructed in such a way that it does not clearly distinguish between the real and the symbolic. This, in turn, means that narrative can play a key role in shaping moral attitudes. In a sense, then, the 'as-if' experience is analogous to the real experience of a situation. It may, therefore, have some contribution to make in shaping our views and guiding our moral choices. Despite its many advantages, the 'as-if' experience is still only a quasi-experience. It can shake the reader, make him or her reflect, but it will never have the same impact as an authentic experience. It gives an impulse for change, but it is an incomplete experience (Treanor 2014, 175). The condition for benefiting from the wisdom contained in a narrative is to make use of it.

6.1.7 Summary

Brian Treanor develops an original conception of virtue inspired by Paul Ricoeur's idea of narrative. The creator of narrative EVE believes that a personal transformation (*metanoia*) of the moral agent can be brought about through an appropriate narrative. A special role in this process is played by moral education, which offers great opportunities for shaping desirable ecological attitudes. However, the shaping of virtues can be guided consciously by the moral agent. The appropriate selection of narratives, mainly from the field of literature, makes an individual's experience a quasi-experience of morality, thereby broadening their own experience and preparing them to make moral decisions in situations analogous to those events learned in the narrative.

6.2 Virtues in a narrative conception of environmental virtue ethics

Treanor notes that our understanding of the word 'virtue' can be distorted for two reasons. First, by often being used in a way that is inconsistent with its original meaning, as happens in the conceptions of contemporary philosophers. Second, advanced theoretical philosophical discussions are too abstract for the average educated person and thus do not serve what Aristotle's virtue served, namely becoming a good person. Treanor emphasizes that it is necessary to define virtue in such a way that the term becomes philosophically precise and meets the theoretical requirements of the discipline. At the same time, it should be formulated in language understandable to the average reader.

Academic philosophy itself faces the serious challenge of not knowing its own history, which results in a misunderstanding of the social and cultural context of the basic concepts used in philosophy. Thus, we use "simulacra" (cf. Baudrillard 2005) of morality, we use terms that function in the tradition, but we do not know their original meaning. We should understand the content of the term 'virtue' and the context in which it was used in ancient culture, and we should be aware of the current social condition and the problems we have to face. Framed in this way, the terminological issues are, on the one hand, grounded in tradition and close to the original understanding; on the other hand, these issues become relevant to the realities of the 21st century. The concept of virtue can then be formulated along the lines of narrative EVE.

6.2.1 A narrative conception of virtue

Treanor, like many other representatives of environmental virtue ethics, adopts the whole range of assumptions of Aristotelian virtue ethics. He understands virtue as a character trait that disposes the moral agent to act in a certain way. A virtuous person acts rightly from the right motives, while feeling the right emotions. Virtue is not something inherent in the moral agent from birth but is something he or she acquires through habituation and sometimes through traditional learning (Treanor 2014, 46). This classical understanding of the concept of virtue should be extended to include the environmental dimension, that is, virtues that improve one's ability to act in a way that responds to the ecological crisis we are experiencing today. As MacIntyre points out, a moral agent does not live in a social vacuum. Environmental virtue ethicists extend this perspective to the entire animate world, pointing out that a fulfilled life cannot take place outside of the environment in which the moral agent lives. It is therefore necessary to develop those character traits that serve the good of nonhuman entities and the human environment.

Environmental virtue serves both a person who possesses it and his environment. Human beings grow thanks to nature (e.g., when one admires the beauty of nature). Nature is valuable to an individual, but at the same time it has intrinsic value. Treanor refers here to Muir's words about the beauty of pristine nature, untouched by man (1901, 4). Moreover, the concept of beauty is one of the key elements of Treanor's understanding of virtue. He uses the Greek term *kalon*,[7] referring to the relationship between goodness and beauty. Indeed, for the Greeks, the good was immanently linked to beauty. Aristotle emphasizes that a virtuous man does beautiful things and derives joy from them. In fact, there is an aesthetic element inherent in the nature of goodness that binds goodness and beauty together in a virtuous man.

The concept of beauty that is woven into the ancient understanding of virtue is the crowning element of Treanor's thought and allows virtue to be defined. Following Sandler, Treanor adopts a naturalistic, teleological, and pluralistic understanding of virtue, to which he adds the aesthetic factor determined by *kalon*, as we come to know it through narratives. Thus,

> a human being is ethically good (i.e., virtuous) in so far as she is well fitted with respect to her (i) emotions, (ii) desires and (iii) actions (from reason and inclination); whether she is thus well fitted is determined by whether these aspects well serve (1) her survival, (2) the continuance of

the species, (3) her characteristic freedom from pain and characteristic enjoyment, (4) the good functioning of her social groups, (5) her autonomy, (6) the accumulation of knowledge, (7) a meaningful life, and (8) the realization of any non-eudaimonistic ends (grounded in non-eudaimonistic goods or values) – in the way characteristic of human beings (i.e., in a way that can rightly be seen as good).

(Sandler 2007, 28)

Treanor add a ninth point to Sandler's definition: "(9) that virtues are traits that have a *kalon* aspect to them and which we understand narratively; therefore, a fully articulated virtue ethics must include a robust narrative component" (Treanor 2014, 50).

6.2.2 Three types of virtue

According to Treanor, virtues relate to three areas of interest to the moral agent: himself, others, and the environment (2014, 55). This approach makes it possible to distinguish three interrelated types of virtues: individual, social, and environmental. There are virtues that are socially relevant and do not affect the environment, such as courtesy. In contrast, the virtue of non-anthropocentrism serves non-eudaimonistic goals, and – as Sandler defines it – is a virtue that serves the environment.

Individual virtues mainly serve the moral agent. These are the virtues that first come to mind when discussing the eudaimonist conception of ethics. As an example, Treanor cites integrity and temperance (2014, 57), a virtue that serves the moral agent to develop restraint and the judicious use of resources, which is very unpopular in consumerist US society. In his reflections, Treanor refers to data on nutrition, which for Americans is very disturbing as one-third of the population is classified as obese. This problem is repeatedly cited in studies of environmental virtue ethics. Emphasizing the environmental costs of resource overexploitation associated with this vice, Philip Cafaro (2005, 140–143) singles out binge eating as one of the four major environmental vices, along with arrogance, greed, and indifference. Treanor puts the problem in other terms by drawing attention to excessive food restriction, which – like intemperance in eating and drinking – is also a vice. Most often, this problem is associated with a serious psychological disorder, such as anorexia.

The consumerist model of life allows material needs to be met relatively easily and quickly. Treanor, citing scientific research (Layard 2005, 1–2), shows that meeting material needs is only to some extent part of the good life. Despite significant improvements in living conditions, people in the United States are no happier than their ancestors were in the 1950s. We have become victims of hedonism, which has a negative impact on ourselves and the environment. We participate in a culture of creating products that are almost disposable; we see constant changes in trends and fashions, and we experience the constant creation of new needs by marketing specialists. All this contributes to a constant sense of dissatisfaction, punctuated by brief periods of fulfilment after acquiring yet another essentially unnecessary product. The compulsion to acquire and buy does not lead to lasting satisfaction and is not a path to happiness in the eudaimonistic sense. Rather, it pushes an individual into a vicious cycle of constantly chasing after new desires. Moderation is necessary, which means proper regulation of one's desires, especially in terms of food, drink, sexual behavior, and so on. Treanor adopts Aristotle's golden mean, which is the path between the two extremes.

The virtue of moderation ensures the physical and mental health of the moral agent by creating the right conditions for virtuous action. Moreover, individual virtues can positively affect the moral agent's environment. The virtue of temperance is a good example of this. Consumerism comes at a huge environmental cost. Markets designed for high consumption are the cause of thoughtless consumption of natural resources. Moderation can have a tremendous impact on the environment and can help curb resource waste. Virtues are present in the culture, but to some extent they have also become part of the private world and the individual aspirations of the moral agent (2014, 20). Treanor criticizes this type of attitude; he believes that making environmental virtues a matter of individual choice regarding lifestyle is a mistake. Environmental virtues should not be solely a matter of personal choice or the pursuit of one's own moral excellence, especially in times of third-generation environmental problems,[8] that is, global environmental problems. Thus, environmental virtues cannot be narrowed down to a private space but are a social issue. The virtue of moderation under discussion also affects the society in which we live as it provides an opportunity to rethink, for example, how we spend our leisure time and choose forms of recreation that do not contribute to excessive resource consumption.

In the group of social virtues, Treanor places courtesy as an example because it is a virtue that primarily serves to build and strengthen constructive social ties. Courtesy is not the only guarantee of good group functioning, but it can significantly improve the way a society functions. It is a typical social virtue that affects the people we surround ourselves with and creates benevolent communities. For Treanor, an example of environmental virtue is holistic thinking, which is directed toward promoting the well-being of the environment. It stands in opposition to the reductionist thinking of the natural sciences. It is an attempt to offer theories that explain all ecosystems. Holistic thinking takes into account the connections between humans and the rest of the ecosystem and concern for the environment and all its elements. Treanor points out that many of the virtues listed in van Wensveen's extensive catalogue relate to holistic thinking, such as ecocentrism and the virtues associated with environmental stewardship. Social virtues help the moral agent to see their own actions in a broader context.

Although Treanor indicates three types of virtues, this does not mean that they are not interrelated. The moral agent functions in each of these three areas, and each virtue can influence the others. Virtues that contribute to the development of the moral agent also benefit his environment in the long term.

Just as social virtues promote the building of healthy communities, they also indirectly influence the moral agent. According to Treanor, the relationship between the three types of virtues can be represented by overlapping circles: each signifies one dimension of virtue functioning, and their overlap delineates a common space that affects more than one plane (Figure 6.1).

The picture shows how the different types of virtues overlap and which of them are in common spaces. According to Treanor, this graphic is an exemplary representation of the complex relationship between different types of virtues. Each of the virtues can interact on surprisingly many levels, and this can often be observed in everyday life. According to Treanor, quite a few virtues affect more than one area of human functioning because individual flourishing is closely linked to a well-functioning social group and environment (2014, 63). An example of a virtue that affects many spheres of human functioning is simplicity, which Treanor discusses in quite some detail. He bases the virtue of simplicity on an analysis of Thoreau's thought and lists its three basic dimensions: professional, material, and intellectual.

Treanor points out that simplicity in Thoreau's terms has three dimensions and refers to the professional, economic, and intellectual spheres. For a discussion of these three spheres, see the chapter on the narrative concept of EVE. Thoreau calls for work to become a vocation and a passion as this constitutes the application of simplicity in the professional sphere. If professional work is just a way to get the money needed to buy more and more goods, it is then worth considering

Virtue

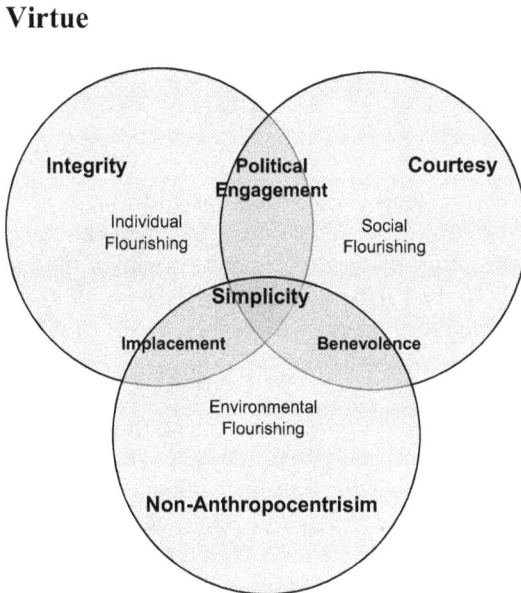

Figure 6.1 Virtues in various dimensions.

Source: Treanor 2014, 61.

whether we are wasting our lives on work (Thoreau 1973, 160). After all, every job must have a clear purpose in sight that is important to us. It must also have a proper place in our lives. Treanor draws attention to workaholism, the bane of today's times, and the fact that we also spend our leisure time on work, such as going to business lunches, checking business emails while on vacation, or never parting with our business phone. This dedication of the entire day to work-related matters is, in his opinion, a dangerous phenomenon.

The second important area in which we should strive for simplicity is the material sphere. Thoreau was an advocate of simple, modest living. His dedication to having few material possessions is evident in the pages of his book that recounts his stay at Walden Lake. He pointed out that owning an excessive number of possessions enslaves a person. In order to understand what we should possess and what we do not need, it is necessary to distinguish between needs and desires. Limiting desires, in Thoreau's view, is a way to increase the quality of life, as we then do not spend our precious time acquiring unnecessary goods and we gain the chance to focus on the quality of life rather than its material aspect. The ascetic life of Thoreau reflects his view on simplicity. From the perspective of life in Western societies, it can be said that he chose ascetic conditions of living. Arguably, such radical principles of limiting the amount of goods would not be met with enthusiasm by the majority of people in Western societies. Treanor (2014, 72) adopts a position similar to Aristotle's golden mean. He criticizes excessive asceticism as well as lack of simplicity.

The third dimension of simplicity relates to the intellectual sphere. Thoreau encouraged people to get rid of everything that is extraneous and superficial in their thinking and focus on what is essential. As he notes, he often found himself dwelling on irrelevant matters, such as street gossip or other trivial news. We often squander precious time and waste our lives pondering things of little importance, allowing

> idle rumors and incidents of the most insignificant kind to intrude on ground which should be sacred to thought. Shall the mind be a public arena, where the affairs of the street and the gossip of the tea-table chiefly are discussed? Or shall it be a quarter of heaven itself – a hypaethral temple, consecrated to the service of gods?
>
> (Thoreau 1973, 171)

We should take a look at our thoughts, for trivial thinking can become a habit into which we will constantly fall; we won't be able to think about great things, and this is part of the good life, that is, living according to high ideals that Thoreau aspired to. Treanor points out that Thoreau's thought is extremely relevant today because we live in a time when a stream of information flows to us from a great variety of sources. Many people are in the habit of constantly checking their mailbox, phone, or social media accounts to read insignificant information. In addition, media outlets heat up emotions so that the news they convey captures our attention. In this way, we become slaves to insignificant issues that it often takes days to ponder. It is therefore necessary to return to simplicity in thinking, to look closely at what we give our attention to and what we feed our thoughts with. This part of living out the virtue of simplicity seems to be the most difficult, as we often don't think about what we are thinking about, spending time pondering things of little importance.

The virtue of simplicity is crucial in all three discussed dimensions, for the moral agent, society, and the environment. In a sense, it is an exemplary virtue that very often appears in environmental discussions as a remedy for the excessive consumerism of the inhabitants of the Earth's northern hemisphere.

6.2.3 The place of virtue in the public sphere

The social dimension is important to Treanor. In his view, environmental virtue ethics is not active enough in the public space. Treanor accuses environmental virtue ethics of lacking what he calls virtue politics (2010, 27), which he understands as collective actions, qualities, or dispositions. These are specific virtues of public life that allow a moral agent to act not only for his own good, but also for the good of the wider community. The lack of virtues in the public sphere – "public virtues" – is one of the charges levelled against environmental virtue ethics.

Virtues are mainly discussed through the lens of individual choice by an individual moral agent, while environmental virtues require the action of a broader group. Environmental challenges will not be solved by the mere involvement of a single person: it is necessary for that individual to become involved in public life and for many moral agents to make a joint effort. Among other reasons, this is because ecological problems involve common goods, which by definition are owned by a wide range of actors. "The problem is that actions that are individually insignificant and therefore not particularly vicious are, when multiplied across a population and combined in effect, both vicious and environmentally devastating" (2010, 16). The tragedy of the common pasture described by Hardin (1968) perfectly illustrates the problem of the conflict between the economic gain of a single entity and the loss of the common good. Treanor cites the analysis of Ostrom (2001), who points out that the sustainable use of a small amount of resources is possible in a sufficiently virtuous community. At the same time, this philosopher points out the need for politicians to work on virtue. They should create laws such that these natural resources are not subjected to loss due to the mechanisms of society. Political virtues must therefore be developed.

Treanor points out that the political dimension of virtues is not a threat to the full development of the moral agent and his pursuit of happiness. Indeed, political virtues are an important element of eudaimonistic ethics (2010, 22). He cites Aristotle, who considers political activity an important part of human life. Being part of the *polis* is one of the most important elements of humans' functioning as it gives them the opportunity to be actively involved in society and provides the space to realize their happiness. In the case of environmental virtues, the area in which the moral agent should be actively involved is the earth system. Treanor points out that there is no way to pursue happiness on a dead planet. Just as personal development is not possible in isolation, achieving happiness is not possible without our contribution and involvement in the community in which we live.

Treanor also points out that the virtues required for political involvement are also good for the development of the moral agent. He claims that political involvement increases the sense of agency. "People engaged in shaping their community and environment feel that they have some influence over their quality of life, which is empowering and satisfying" (2010, 22). There are many virtues that influence a moral agent's involvement in pro-environmental or political activities, and they all contribute to his human flourishing.

6.2.4 Summary

Treanor criticized the theoretical dimension of environmental virtue ethics. The accusation he levelled against ethicists is that their analyses end only in book publications. In his opinion, environmental virtues should be implemented on a large scale because they concern three spheres of human functioning: oneself, others (the social dimension), and the environment. When talking about environmental virtues, the discourse should not be narrowed down to nature alone: the personal development of the moral agent should be kept in mind, as well as virtues that affect the moral agent's relations with others.

Treanor introduces a new category, narrative, into the discussion of environmental virtues. He also draws attention to the role of metanarratives (e.g., religion), literature, and self-narratives in shaping moral attitudes. Although the very concept of narrative has already appeared in earlier philosophical discussions, it is new in environmental virtue ethics. Narrative is a very useful tool for at least several reasons. First, it provides a subtle way of conveying information regarding virtues, avoiding a moralizing tone. Second, the metaphors implicit in a narrative trigger the desire to strive for ideals. Although they are subtle, they provide a very clear and explicit incentive to develop the potential to do good. Third, narrative provides a kind of experience, thus allowing the moral agent to gain the knowledge and practice necessary for the formation of moral attitudes. Fourth, narrative is an extremely important tool for facilitating self-understanding. Through self-narration, we can interpret and understand our own experiences. The narrative concept of environmental virtue ethics allows us to better grasp the phenomenon of environmental virtues. As Gammon notices, Treanor's unification of environmental, virtue, and narrative ethics ultimately presents a persuasive case for environmental thinkers to embrace the offerings of virtue ethics and to remember the wealth of environmental narratives from which we can already draw inspiring and edifying lessons about the good life (2016, 381).

Notes

1 Brian Treanor is a Fellow at Loyola Marymount University, where he is director of the Academy for Catholic Thought and Creativity, an interdisciplinary research center that encourages intellectual dialogue at the university and supports research in the humanities, social sciences, arts, and sciences. He received his undergraduate degree from the University of California, after which he continued his education at California State University, Long Beach, and later at Boston College. Brian Treanor is an avid traveler, having visited six continents and lived on three (source: http://faculty.lmu.edu/briantreanor/about-2).

2 The concept of narrative was not created by Treanor: it was already present in various forms in environmental philosophy; he just applied it to environmental virtue ethics, which makes his approach unique in terms of EVE. Ryan claims that "over the last 30 years, environmental philosophers and ecological researchers have debated the potential of narratives for generating 'relational ethics' (Ellis 2007), place-based ethics (Friskics 2023), and non-anthropocentric models of ecological intervention (see Cheney 1987; Warren 1993; Preston 2001; Robertson et al. 2001; Liszka 2003; Slicer 2003)" (2012, 822). This plethora of perspectives on narrative approaches in environmental philosophy shows the importance of the proper usage of human–nature narratives.

3 Priming: "in cognitive psychology, the effect in which recent experience of a stimulus facilitates or inhibits later processing of the same or a similar stimulus. In repetition priming, presentation of a particular sensory stimulus increases the likelihood that participants will identify the same or a similar stimulus later in the test. In semantic priming, presentation of a word or sign influences the way in which participants interpret a subsequent word or sign" (APA 2018). When discussing priming and other psychological concepts, one should be aware of the replicability crisis, that is, the issue that many scientific studies prove difficult or impossible to reproduce. It was John Bargh's experiment on "elderly walking" (1996) that led to the concept of priming but attempts to replicate it failed. Recognition of the replicability crisis has led to discussion and thus to the elaboration of more precise and credible methodology. Thus, all references to psychological research in this book are supplementary to the philosophical discussion (cf. Bargh et al. 1996; Doyen et al. 2012; Wiggins, Christopherson 2019; Maxwell et al. 2015).

4 The Jataka is a collection of 574 stories about Buddha incarnations. They are considered one of the most important subjects of Buddhist art in India and abroad.

5 The most famous collection of Hindu fables, written in India between the 4th and 3rd centuries BC.

6 Stegner (2007) is the author of a well-known letter in the environmental literature in defense of pristine nature.

7 *Kalon* is an ideal of physical and moral beauty characteristic of ancient Greek philosophy. It refers to the concept of *kalokagathia* (καλοκαγαθία), or the belief in the inseparability of beauty (*kalos*) and goodness (*agatos*). This concept is evident in Aristotle's understanding of virtue, as well as in Plato's Dialogues.

8 According to Ronald Sandler, in the first generation it was important to protect the environment somewhere "out there," such as the protection of endangered species or the debate over building roads in protected areas. The second generation brought problems that are "right here," such as exhaust fumes or pollution of other environmental resources. The third generation brought problems that are "everywhere," such as climate change or dwindling resources. "Everywhere" from the third generation was external, and in the fourth generation it will be a very personal and internal factor. It will refer to the problems generated by genetic engineering and nanotechnology (Sandler 2007, 3).

References

Abbey E., *Desert Solitaire*, New York 1985.
APA Dictionary of Psychology, Priming, 2018, https://dictionary.apa.org/priming, (Access: November 30, 2023).
Bargh J.A., Chen M., Burrows L., *Automaticity of Social Behavior: Direct Effect of Trait Construct and Stereotype Activation on Action*, "Journal of Personality and Social Psychology" 1996, vol. 71, no. 2, p. 230–244.
Baudrillard J., *Symulakry i symulacja*, transl. S. Królak, Warsaw 2005.
Brower D., *Let the Mountains Talk, Let the Rivers Run: A Call to Those Who Would Save the Earth*, Gabriola Island B.C. 2000.
Cafaro P., *Gluttony, Arrogance, Greed, and Apathy*, in: *Environmental Virtue Ethics*, R. Sandler, P. Cafaro (ed.), New York 2005, p. 135–158.
Caputo J. D., *More Radical Hermeneutics*, Indianapolis 2000.
Cheney J., *Eco-feminism and Deep Ecology*, "Environmental Ethics" 1987, vol. 9, no. 2, p. 115–145.
Cielemęcka O., *Perspektywa zdolności a sprawiedliwość społeczna. Recenzja: Martha C. Nussbaum, Creating Capabilities: Human Development Approach, Belknap Press of Harvard University Press, 2011*, "Etyka" 2011, vol. 44, p. 183–186.
Doyen S., Klein O., Pichon C. L., Cleeremans A., *Behavioral Priming: It's all in the Mind, But Whose Mind?*, "PLOS ONE" 2012, vol. 7, no. 1, DOI: 10.1371/journal. pone.0029081
Dzwonkowska D., *Normatywność etyki cnót środowiskowych na przykładzie etyki Ronalda Sandlera. Komentarz*, "Avant" 2017, vol. 8, no. 3, p. 99–108.
Ellis C., *Telling Secrets, Revealing Lives: Relational Ethics in Research with Intimate Others*, "Qualitative Inquiry" 2007, vol. 13, no. 1, p. 3–29.
Friskics S., *Dialogue, Responsibility, and Oil and Gas Leasing on Montana's Rocky Mountain Front*, "Ethics & the Environment" 2023, vol. 8, no. 2, p. 8–30.
Gammon A. R., *Emplotting Virtue: A Narrative Approach to Environmental Virtue Ethics*, "Environmental Ethics" 2016, vol. 38, no. 3, p. 379–382.
Gergen M. M., K. J. Gergen, *What Is This Thing Called Love? Emotional Scenarios in Historical Perspective*, "Journal of Narrative and Life History" 1995, vol. 5, no. 3, p. 221–237.
Goleman D., *Inteligencja ekologiczna*, tłum. A. Jankowski, Poznań 2009.
Hardin G., *The Tragedy of the Commons*, "Science" 1968, vol. 162, no. 3859, p. 1243–1248.
Hardy B., *Towards a Poetics of Fiction: An Approach through Narrative*, "NOVEL: A Forum on Fiction" 1968, vol. 2, no. 1, p. 5–14.
Jaśtal J., *Spór "fakultetów" o pojęcie cnoty. Konstruktywistyczny dyskurs etyczny wobec dyskursu naukowego*, in: *W poszukiwaniu moralnego charakteru*, N. Szutta, A. Szutta (red.), Lublin 2015, p. 259–289.
Jayawardena L., *Foreword*, in: *Women and Human Development: The Capabilities Approach*, M. Nussbaum (ed.), Cambridge 2001, p. V–VI.
Kearney R., *Ethics of the Narrative Self*, in: *Between Philosophy and Poetry: Writing, Rhythm, History*, M. Verdicchio, R. Burch (ed.), London 2002.
Krakauer J., *Into the Wild*, New York 1997.
Layard R., *Happiness: Lessons from a New Science*, New York 2005.
Liszka J., *The Narrative Ethics of Leopold's Sand County Almanac*, "Ethics & the Environment" 2003, vol. 8, no. 2, p. 42–70.
Maxwell S. E., M. Y. Lau, G. S. Howard, *Is Psychology Suffering from a Replication Crisis? What Does "Failure to Replicate" Really Mean?*, "American Psychologist" 2015, vol. 70, no. 6, p. 487–498.
McIntyre A., *After Virtue. A Study in Moral Theory*, Notre Dame 1988.

Muir J., *Our National Parks*, New York 1901.

Muir J., *My First Summer in the Sierra*, Mineola NY 2005.

Nussbaum M., *Form and Content, Philosophy and Literature*, in: eadem, *Love's Knowledge. Essays on Philosophy and Literature*, Oxford 1990.

Nussbaum M., *Women and Human Development: The Capabilities Approach*, Cambridge 2001.

Ostrom E., *Reformulating the Commons*, in: *Protecting the Commons: A Framework for Resource Management in the Americas*, J. Burger, E. Ostrom, R. B. Norgaard, D. Policansky, B. D. Goldstein (ed.), Washington 2001.

Peacock D., *Grizzly Years: In Search of the American Wilderness*, New York 1996.

Powell R. C., *Transforming Genius into Practical Power. Muir, Emerson, and the Politics of Character*, "Environmental Ethics" 2020, vol. 42, no. 1, p. 21–37.

Preston Ch. J., *Intrinsic Value and Care: Making Connections through Ecological Narratives*, "Environmental Values" 2001, vol. 10, no. 2, p. 243–264.

Prior W.J., *Eudaimonism and Virtue*, "The Journal of Value Inquiry" 2001, vol. 35, no. 3, p. 325–342.

Ricoeur P., *Time and Narrative*, vol. 1, transl. K. McLaughlin, D. Pellauer, Chicago 1984.

Ricoeur P., *Oneself as Another*, transl. K. Blamey, Chicago 1992.

Robertson M., Nichols P., Horwitz P., Bradby K., MacKintosh D., *Environmental Narratives and the Need for Multiple Perspectives to Restore Degraded Landscapes in Australia*, "Ecosystem Health" 2001, vol. 6, no. 2, p. 119–133.

Rosner K., *Paul Ricoeur wobec współczesnych dyskusji o narracji*, "Teksty Drugie" 2002, vol. 22, no. 3, p. 129–136.

Ryan J. C., *Narrative Environmental Ethics, Nature Writing, and Ecological Science as Tradition: Towards a Sponsoring Ground of Concern*, "Philosophy Study" 2012, vol. 2, no. 11, p. 822–834.

Sandler R., *Character and Environment. A Virtue-Oriented Approach to Environmental Ethics*, New York 2007.

Sapolsky R., *This is Your Brain on Metaphors*, "New York Times", 2010, https://archive.nytimes.com/opinionator.blogs.nytimes.com/2010/11/14/this-is-your-brain-on-metaphors/ (Access: June 12, 2022).

Slicer D., *Introduction: Special Issue on Environmental Narrative*, "Ethics & the Environment" 2003, vol. 23, no. 1, p. 1–7.

Stegner W., *To: David Pesonen, December 3, 1960*, in: *The Selected Letters of Wallace Stegner*, P. Stegner (ed.), Emeryville, CA 2007, p. 352–357.

Szutta N., *Dyskusja z sytuacjonistyczną krytyką etyki cnót. Odpowiedź na zarzuty Gilberta Harmana*, "Diametros" 2012, vol. 9, no. 31, p. 88–112.

Szutta N., *Wprowadzenie, czyli o co chodzi w dyskusji nad moralnym charakterem*, in: *W poszukiwaniu moralnego charakteru*, N. Szutta, A. Szutta (ed.), Lublin 2015, p. 13–34.

Thoreau H.D., *Life Without Principle*, in: *The Higher Law*, W. Glick (ed.), Princeton 1973.

Treanor B., *Narrative Environmental Virtue Ethics: Phronesis without a Phronimos*, "Environmental Ethics" 2008, vol. 30, no. 4, p. 361–379.

Treanor B., *Environmentalism and Public Virtue*, in: *Virtue Ethics and the Environment*, P. Cafaro, R. Sandler (ed.), Dordrecht 2010, p. 9–28.

Treanor B., *Emplotting Virtue: A Narrative Approach to Environmental Virtue Ethics*, Albany 2014.

Treanor B., *Vitality. Carnal, Seraphic Bodies*, "Journal of French and Francophone Philosophy – Revue de la Philosophie Française et de Langue Française" 2017, vol. 25, no. 1, p. 200–220.

Warren K., *The Power and Promise of Ecological Feminism*, in: *Environmental Philosophy: From Animal Rights to Radical Ecology*, M. Zimmerman et al. (ed.), Englewood Cliffs 1993, p. 320–337.

Wiggins B. J., Christopherson C. D., *The Replication Crisis in Psychology: An Overview for Theoretical and Philosophical Psychology*, "*Journal of Theoretical and Philosophical Psychology*" 2019, vol. 39, no. 4, p. 202–217.

Wolicka E., *Mimetyka i mitologia Platona*, Lublin 1994.

Part III

Presentation of the universalistic, positive, and practical environmental virtue ethics

This section of the monograph is devoted to presenting the universalist, positive, and practical concept of environmental virtue ethics. In my view, these features of environmental virtues are crucial in enabling the application of environmental virtues. Since they have not yet been addressed in the discussion, I decided to take a closer look at them. Due to the universal nature of the environmental crisis, it is possible to develop catalogues of environmental virtues that apply to diverse societies and constitute a factor that unites them. Thus, in a time when individuality, separateness, and differences are emphasized, the universalist factor could be crucial in overcoming the environmental crisis. Knowledge of the positive nature of virtues allows better selection of narratives used in environmental protection. Knowing that negative information doesn't work (or works only to a limited extent) and that so-called "ecologies of fear" have long been discredited, we can harness the positive message provided by virtues to bring about changes in the attitudes of moral agents. The practical dimension, on the other hand, refers to the realm of praxis, for virtue is never just a theoretical concept. This element concerning virtue has appeared in previous discussions, but I nevertheless discuss it from a slightly different angle.

7 Evaluation of the three concepts of environmental virtue ethics (EVE)

The beginning of this section of the monograph is devoted to an evaluation of the three existing EVE concepts. The discussion so far on the ethics of environmental virtues has significantly enriched the debate on humans' ethical obligations to the environment. First of all, it has shown the relationship between the character traits of the moral agent and concern for the natural environment, which is a significant contribution to the development of analyses concerning the moral sphere of the human–environment relationship. Secondly, each concept of EVE contributes a brick to the creation of catalogues of virtues and to the theory of environmental ethics as each one contains interesting views on environmental virtues and contributes to the construction of an environmental ethos, creating an aretological landscape of environmental virtue ethics in American philosophical literature. Some critical remarks that might appear in this chapter are in no way intended to discredit these very important, valuable, and impactful concepts of EVE.

7.1 Evaluation of the classic concept of EVE

The classic concept, because of the figure of Thoreau, is the most popular concept of EVE. Thoreau is undoubtfully a hero of the United States's environmental movement and pop culture. His writings, attitude, and life, despite the passage of years, continue to inspire generations of environmentalists. It is a concept that from beginning to end is a description of his own experience. Thus, it is so authentic that, despite the passage of time, *Walden; or, life in the woods* is a source of inspiration. Moreover, the power of Thoreau's myth makes his figure and activities widely known. Thoreau's concept, although it may rank as a classic, is not without its imperfections.

The first objection that can be levied against the theoretical side of Thoreau's concept stems from his lack of philosophical background. His book about his sojourn at Lake Walden, although it contains important and philosophy-bearing themes, is more a literary work than a philosophical one, which makes it more of a background for later philosophical analysis rather than a philosophical work itself. Thoreau's book is far from perfect in terms of theoretical

DOI: 10.4324/9781003433156-11

elaboration itself. Here, the later interpretation of Thoreau's thought, of which the most comprehensive is by Cafaro (2004), gave Thoreau's ideas a somewhat more philosophical dimension and has undoubtedly proved valuable. Nevertheless, the role that Thoreau plays in discussions of environmental ethics is of such significance that omitting it in this book would give an unreliable account of the debate over environmental virtue ethics. After all, Thoreau himself is a hugely influential role model and a key figure for many generations of environmentalists.

Thoreau chose the path of individual asceticism against the currents that were typical of Transcendentalism, which offered the possibility of putting philosophical ideas into practice by living in communities. This attitude on the part of Thoreau, probably stemming from his aversion to maintaining social relationships, raises some questions. Certainly, for a good life it is crucial to develop one's inner self; nevertheless, we can often polish our character and cultivate virtues only through contact with other people. Interpersonal relations allow us to develop the ability to respond adequately to a given situation, to empathize with the other, to read the other's need. Social relations expose us to moral challenges that we do not experience when living in isolation. When isolated from society, a person does not have the opportunity to learn about himself and see if he indeed has moral virtues. How, for example, can one know that one will not lie if one is not put to the test by another person's questions?

Negligence of the social sphere and excessive focus on relations with the natural environment are often found in environmental ethics. Environmental virtues refer to our functioning in the natural environment, but humans are beings whose existence depends not only on the natural world, but also to a large extent on the social context. MacIntyre even writes about the communitarian dimension of virtue. He points out that virtues are always developed and practiced in some community, which is important for the development of the moral subject. Holmes Rolston III (2005) goes even further and defines environmental virtues detached from their social context as a half-truth that is dangerous to integral human development. He emphasizes the fact that humans must develop not only environmental virtues but also social virtues that regulate interpersonal relations. Thus, Thoreau's ideal refers only to a certain sphere of human life, and it would be appropriate to expand his proposal to include the set of social virtues necessary for the moral subject to function in the environment.

One could ask about the relevance of Thoreau's attitude to modern times. As Treanor points out, the modern reader will at best think of this American transcendentalist as out of touch, and at worst as a madman (cf. Treanor 2007, 65). Thoreau's experiment looks like a bold if not downright crazy move, to say the least. It definitely doesn't fit in with a consumerist society that is unlikely to care about saving resources. Although Treanor points out that viewing Thoreau as a madman demonstrates a lack of understanding of his ideas, in my opinion it is important to look at the idea of simplicity that is promoted in his work.

The previous chapters of this book have described Thoreau's understanding of simplicity, but in this part I will focus primarily on simplicity understood as leading a modest, simple, even ascetic life.

It is this extreme asceticism that seems most problematic here, for it is not in the least compatible with the consumerism of today. Descriptions of a very modest and simple life may impress but will never become the standard for most people. Moreover, Thoreau's understanding of simplicity has nothing to do with Aristotle's view of virtue as a golden mean between excess and deficiency (2007, II, 7), where both extremes are vices. According to Aristotle, every virtue has two poles that are wrong and do not serve the moral subject. According to Aristotle,

> with regard to feelings of fear and confidence, courage is the mean; of the people who exceed, he who exceeds in fearlessness has no name (...), while the man who exceeds in confidence is rash, and who exceeds in fear and falls in confidence is a coward.
>
> (2007)

With regard to temperance, both excessive pleasure seeking and excessive indulgence in pleasures can be considered vices. The other extreme – a person deficient of pleasures, as Aristotle writes – is not often found (2007). Thoreau's asceticism would be such a huge sacrifice for modern members of Western societies that it wouldn't fit into Aristotle's understanding of virtue as a golden mean between two extremes. Thus, would Thoreau's asceticism, according to Aristotle, be a vice? Probably not, but it is a supererogatory character trait, therefore only a few members of society can live up to this standard.

An interesting opinion on this can be found in Treanor's writing (2007, 81), according to which the virtue of simplicity should not be viewed in terms of the golden mean because its essence is the renunciation of excess goods. For this reason, it is rather closer to one of the extremes. The very nature of the virtue of moderation implies a kind of "condemnation to mortification." Regarding simplicity, Treanor even writes that one should be wary of pleasures and take pleasure in simplicity (2007, 82). This position is not far from Aristotle's ideal, for the Stagirite notes that "one extreme is (always) nearer and liker to the intermediate" (2007). Thus, one of the extremes seems to be better than the other. Thus, Thoreau's moderation seems to be more virtuous than indulgent. Thus, maybe the closeness of asceticism to the golden mean would be enough to consider Thoreau's moderation as a virtue. After all, the extent to which one is able to apply moderation is very individual. The problem in leading an ethically good life stems from the fact that

> it is no easy task to find the middle; e. g., to find the middle of a circle is not for everyone but for him who knows; so too, anyone can get angry – that is easy – or give or spend money; but to do this to the right person,

to the right extent, at the right time, with the right motive, and in the right way, that is not for everyone, nor is it easy; wherefore goodness is both rare and laudable and noble.

(Aristotle 2007, II, 9)

Finding the right measure is a challenge for ethics and a task for the moral agent. Therefore, with regard to environmental virtues, Treanor advocates taking individual predispositions into account. This is an argument that opens up the possibility of viewing Thoreau's attitude in terms of virtue. First, at the time of his stay at Lake Walden, living standards were much lower than they are today in highly developed societies. Second, he probably did not see the rejection of the goods of civilization as an excessive sacrifice. A reading of his book shows that he was very focused on perfecting his character and saw simple living as a tool for achieving this goal; so, for him, modest living conditions were not an expression of extreme mortification. Third, it is clear from his descriptions that the idea of gaining freedom by not having many possessions was so important to him that he derived satisfaction from his ascetic life.

Thus, given Treanor's suggestion that virtue should be adapted to the individual capabilities of the moral agent, the model of simple living advocated by Thoreau can be considered virtuous only by some. It is incompatible with the current realities of consumer society. However, the very idea of such radical asceticism should not be considered an unhelpful guideline in today's world. Thoreau's attitude does not lend itself to being copied by everyone; it sets a goal or provides inspiration, rather than providing a ready-to-apply template embedded in an ethic of environmental virtue. A certain cautious skepticism should be adopted toward the ideals presented by Thoreau, since blind imitation could discourage environmental protection or even have negative consequences, as shown by the example of McCandless, whose life ended tragically.

Thoreau's attitude and his understanding of simplicity could be considered heroic virtues, and the actions resulting from following these virtues would be an example of supererogation. However, it should be remembered that not everyone can implement heroic virtues, so the model of modesty proposed by Thoreau is not one that can be universally applied in modern consumer societies. Such an ascetic ideal of modesty undermines the value of Thoreau's works and puts the achievement of virtue beyond the reach of a sizable portion of the population of Thoreau's homeland.

Coeckelbergh (2015, 4) notes that Thoreau's attitude is incompatible with the realities of environmental protection because it is based on a romantic perception of nature, which Coeckelbergh considers inappropriate. Indeed, from Thoreau's writings shines through the myth of an uncreated, infinitely good nature, contrasted with imperfect civilization and culture. According to Coeckelbergh, the biggest problem is the two opposing spheres: the world of good nature and bad civilization. In my opinion, however, the biggest problem is a naive and unrealistic view of nature. This type of thinking has at least two major flaws. First, it does not allow one to read reality as it is. At best, this approach can be treated as a literary description, not a philosophical reflection. Second, it can have serious consequences for potential imitators. The case of McCandless is tragic proof of this.

7.1.1 Summary

Undoubtedly, Thoreau cannot be denied an important place in American culture. He is an icon of the environmental movement and an unsurpassed, even heroic, example of the realization of environmental virtues. No book on environmental philosophy in international literature can be complete without reference to the person of Thoreau. Admittedly, it must be acknowledged that, as is sometimes the case with pop culture icons, Thoreau's thought has also been somewhat exaggerated and distorted. The biggest challenge with Thoreau's philosophy is its incompatibility with daily practice. The ecological ethos established by Thoreau is a supererogatory model for today's consumer societies, so for many people it is virtually impossible to adopt such a radical attitude. It has served as an inspirational vision but it could be implemented by only a few very determined individuals.

7.2 Evaluation of the naturalistic, teleological, and pluralistic conception of EVE

Sandler's concept is a very valuable voice in the aretological discussion. Sandler presents his position on EVE while referring to contemporary concepts of virtue ethics, namely the views of Rosalind Hursthouse and Philippa Foot, both of whom are among the most recognizable and influential figures in the contemporary renaissance of virtue ethics. Thus, from a perspective broader than environmentalism, it is important to show the applicability of virtue ethics to one of the most serious problems of contemporary civilization. Sandler's concept is an original way of approaching the environmental question, for it is rare in the environmental ethics literature to refer to mainstream philosophical discussions to such a wide extent. Often, environmental philosophers operate in the hermetic language of their discipline, referring to other thinkers dealing with the environmental question. This reference to contemporary philosophy gives Sandler's concept a very solid theoretical foundation. The strong theoretical underpinning of his approach is an asset of the naturalistic, teleological, and pluralistic conception of EVE.

Even though Sandler's concept has a very well-prepared theoretical background, some concepts seem to apply to a moral agent who is more proficient in terms of his moral dispositions, for example the extension strategy (Sandler 2005, 4), according to which one can extend virtue to the environmental dimension of human functioning. Thus, a question arises: Does every human have such proficiency in analyzing moral situations that they would be able to make a proper extension of virtue to the natural environment? An example of a poorly founded extension of virtue is the idea of friendship with the environment, understood as environmental virtue (Cf. Bannon 2017). The very idea of friendship with non-personal entities requires a change in the way friendship is defined, namely as an interpersonal relationship. It is also necessary to redefine the question of reciprocity in friendship. These changes in the way the concept of friendship is understood prompt some questions: Can we still speak of friendship in such a changed conception of this virtue? Is the use of the word 'friendship' legitimate? Are we not, by any chance, confusing the content of the terms used? It should be emphasized that Ronald Sandler criticized the concept of friendship with the Environment. However, the extension of virtue seems to need prudence for a moral agent to know how and when to apply it.

One of the strong aspects of Sandler's concept is the embedding of his theory in naturalism. The concept of the naturalistic approach seems to be very relevant and needed in contemporary ethical discussions as it provides a holistic picture of a human being as an entity that is part of nature and part of the world of culture, that is, an entity determined by his own biology but having the potential to transcend it and the ability to control his own conditioning or consciously choose which biological impulses to succumb to. In European culture, attitudes toward the body and corporeality are often the result of falling into one of two extremes, that is, various forms of fetishization of the body (e.g., in hedonistic currents or contemporary consumerism), or various forms of depreciation of corporeality (e.g., Gnosticism, mainstream medieval thought (cf. MacIntyre 1996, 300), Cartesian dualism). The lack of recognition of corporeality as an integral part of human beings was strongly influenced by medieval thought, which sought to create a philosophy directed toward goals higher than satisfying hedonistic needs. In this perspective, the body became less important, and sometimes even the desires arising from corporeality became an obstacle to the realization of higher – in this case religious – goals. The philosophical perception of the body and corporeality has become the subject of numerous studies; however, as this is only a side thread, in this monograph I will focus on a very perfunctory discussion of philosophers' views on this topic. Among other things, I will stop at medieval thought because its influence on European culture is often underestimated, whereas it represents such a significant time slice of our tradition of philosophical reflection that it has left a substantial mark on European culture.

Dorta Zydorek (1996) points out that the thought of Saint Augustine influenced attitudes to corporeality in the Middle Ages. The Bishop of Hippo preached that man consists of two natures, spiritual and corporeal (cf. Gilson 1953, 61), and that the body is, on the one hand, an integral part of the personal being and, on the other, it repeatedly stands in the way of salvation. This view ran through medieval philosophy in various forms. According to Zydorek, the essence of medieval views on corporeality was laid out in Pope Innocent III's 12th-century treatise *De contempt mundi sive de miseria conditionis humanae libri tres* (1828). He depicts the human body as being made of mud, ash, and the foulest semen, emphasizing that human beings, by means of asceticism, should free the soul from its scarlet bodily garments. These rather radical descriptions correlate with the mainstream of medieval thought, which tended to pursue higher goals of religious life, while the body was perceived as an obstacle to sanctity. Hence, various forms of asceticism emerged at that time that sought to control and tame the body.

Ten centuries of medieval thought (and the transitional period between antiquity and the Middle Ages) left their mark on Western philosophy, which did not pay excessive attention to corporeality or the biological aspect of human nature. As a consequence, our ignorance of human nature became apparent and transferred to the natural world as a whole. Only occasionally did various kinds of hedonistic views emerge which brought elements of carnality into culture, albeit in an extreme form. This legacy was legitimized in culture through a number of later views (e.g., Cartesian dualism or the views of F. Bacon). Although many philosophers have recognized a naturalistic element in environmental philosophy, the nature–culture dualism has contributed significantly to the contemporary environmental crisis.

In my view, the naturalistic element in Ronald Sandler's thought is an element that ties the high ideals of virtue ethics to the biology of human beings. Sandler's perspective is closer to Aristotle's view of virtues as the golden mean between two extremes. It is also closer to the ancient way of seeing virtues as a kind of disposition that could be seen in any activity and in the performance of everyday duties. Above all, however, Sandler's perspective makes virtue something real that is not an act of heroic asceticism but is related to the everyday life of the moral agent. A proper grasp of the body and an understanding of its role helps the moral subject assess their own capacities, including how best to select the virtues that need perfecting and how to skillfully put them into practice.

However, it should be stressed that a superficial perusal of Sandler's conception may lead to the accusation that he commits a naturalistic error, in this case understood as identifying the term 'good' in the ethical sense as a natural quality (cf. Jedynak 1967, 290). However, it should be emphasized that the naturalistic understanding of the good in this view is intended to show the relationship between a human being and their corporeality. Biological factors are not presented from a perspective that suggests their equivalence with, or the possibility of replacing, the content of the term 'good' in the ethical sense. Furthermore, the word 'good' in relation to nonhuman entities or the natural world is used not with an ethical meaning but in a colloquial sense. Thus, it has been used synonymously with the term 'well-being,' but not in relation to goodness as an ethical category.

7.2.1 Summary

In summary, Sandler's concept has undoubted theoretical merits. These include, first and foremost, the strong embedding of EVE theory in the theory of contemporary virtue ethics. Its second great value is its emphasis on the biological and cultural dimensions of humanity. Many philosophical concepts seem to forget the cultural aspect, deriving reductionist philosophical theses from empirical research; in turn, some strands of contemporary philosophy are silent on the biological dimension. By emphasizing that man is both a biological and a cultural entity, it is possible to look at his nature in a way that is adequate to his essence.

7.3 Evaluation of the narrative conception of EVE

The narrative concept of EVE is very much focused on the practical dimension. What is important is the application of the virtues shown through an appropriate narrative or used as a political tool. This is, in my opinion, the most important element of Treanor's thought. I believe that narrative as a tool for talking about virtues does an excellent job of allowing moral virtues to be reinforced in a subtle yet effective way. Narrative conveys good content through images, which is a more effective way than other forms of communication. What's more, it is a tool adapted to people of all ages and levels, each of whom can choose the right narrative for I and their needs.

Although Treanor acknowledges narrative as a great tool for transmitting and teaching virtues, it has to be accompanied by a properly designed moral education. As Julia Annas writes,

> We don't learn to be virtuous from books, even books about virtue, though these help our understanding. Aristotle is right here: virtue is like building in that learning to be brave is learning to do something, to act in certain ways, and that where we have to learn to do something, we learn it by doing it (not just by reading books about it).
>
> (2011, 22)

In Julia Annas's view, virtues are learned in the concrete situations in which we find ourselves. We learn virtues at school, work, and church and from parents, teachers, friends, neighbors, and the internet. As Annas points out, in the early stages of development, in childhood, imitation of role models[1] is crucial and is initially completely uncritical and unconscious. Over time, the moral agent begins to mature and to reflect on their own behavior, thus thinking more and more consciously about their actions.

Annas also notes that certain cultural patterns transmit some sort of image of virtue. In environmental discussions, this issue has been raised by van Wensveen (2000, 132–135), who claims that role models for courage are built on a "Rambo-like picture" (2000, 132). This Dutch philosopher expresses concern that this image of courage does not allow the feminine aspect of virtue to be expressed, and the character trait itself is presented in such a limited way that does not express its richness. Even though in contemporary times we might discuss whether the feminine aspect of this or other virtues is visible, this concern enables one to reflect on the role of popular culture in character formation.

One can agree with the thesis that we are caught up in understanding virtues from the perspective of the culture in which we were raised or in which we reside, or one with which we identify. Given the process of globalization, we are extensively exposed to the subtle influence of cultural patterns promoted by the mass media. Therefore, the moral agent should consciously analyze the images they see in order to understand virtues, and not just perceive the patterns impontempy popular culture. In line with Annas's and van Wensveen's comments, we should ask whether literature is enough to shape new moral examples. Indeed, literature can be very helpful in shaping moral skills, but its role cannot be depreciated. In times of mass media dominance, the role of literature is not as significant as in earlier eras. Today's elites, unlike those of earlier times, do not meet to read various works of literature. Reading is very much on the decline and the role of short messages conveyed by various types of mass media on the internet is growing. Thus, the aretological message should not only flow from literary works but should also be disseminated through the various media available. The question is not if narrative ethics could support moral education, but how it could support it. Maybe moral education is a process that is already happening in media presenting various dimensions of the environmental crisis and ways of reacting to it. If the process of moral education has started, the only missing element would be to make positive environmental discourse a mainstream media topic.

Narrative is a promising tool in the field of ethical formation that allows us to escape from moralistic formulae and use our human capacity to think in metaphors and images. Narrative sets the framework for putting ethics into practice. It is important to adapt the mode of narrative to the historical and social context of moral subjects exposed to ethical formation by means of narrative, and the selection of appropriate literature is crucial. Treanor has highlighted this issue by presenting the figures of Don Quixote and McCandless. Undoubtedly, both characters had noble motivations, but motivations themselves do not exemplify virtue. First, both of them showed a lack of *phronesis* in assessing their own situations and choosing appropriate actions. Prudence is one of the key virtues as it is what enables a moral agent to know when and to what extent to apply a particular virtue. It has the function of integrating and controlling the action of the moral subject, helping to determine the right measure – a middle way between two extremes. Undoubtedly, Don Quixote and McCandless are not good examples of following the path of virtue rationally. In his book (2014), Treanor has shown the importance of proper selection of role models and literature.

The narrative concept of EVE forces one to ask who would choose the narrative for the moral agent: would it be the moral agent himself who would choose his readings? Such a position is fraught with error, as Treanor points out. Each of us has our own preferences and tastes. The moral agent may not be interested in the patterns that lead to the formation of environmental virtues. Various individuals may have different interests or even prefer narratives that do not serve to promote any positive role models. Individual preferences condemn us to resort to narratives that suit our tastes, so the key question is how to ensure that the narratives we choose offset the problem of deficits in environmental morality. Moreover, mainstream literature is not much concerned with environmental problems.

Despite the preceding reservations, I believe that narrative is a good auxiliary tool for the formation of virtue. Of course, the way it is used needs to be broadened (i.e., supplemented with mediums other than just literature and the use of mass media and social media). Moreover, we need to reflect on ways in which we present moral exemplars. The great values of narrative are its lightness, its subtlety, and its great transformative potential. Hence, in my view, Treanor's idea of integrating narrative into environmental virtue ethics is extremely valuable as it promotes goodness subtly but effectively, allowing us to escape the moralistic tone of talking about virtue. When used consciously, narrative can prove to be one of the most effective tools in shaping virtues.

7.3.1 Summary

The narrative conception of environmental virtue ethics is a very interesting account that has many good points. First, due to Treanor's broad intellectual horizons and erudition, just reading his paper is extremely engaging and one can benefit greatly from reading his work. Second, narrative is, in my opinion, the best practical tool proposed so far in environmental virtue ethics, providing an opportunity to talk about virtue in a subtle and non-imposing way.

Note

1 It should be emphasized that the issue of environmental role models is quite well covered in the literature (cf. Cafaro 2011; Sandler 2005); also, Treanor (2014) covers this issue, explaining in detail his perspective on environmental heroes.

References

Annas J., *Intelligent Virtue*, Oxford 2011.
Aristotle, *Nicomachean Ethics*, transl. W. D. Ross, 2007, https://socialsciences.mcmaster.ca/econ/ugcm/3ll3/aristotle/Ethics.pdf (Access: October 15, 2022).
Bannon B. E., *Being a Friend to Nature: Environmental Virtues and Ethical Ideals*, "Ethics, Policy & Environment" 2017, vol. 20, no. 1, p. 44–58.
Cafaro P., *Thoreau's Living Ethics. Walden and the Pursuit of Virtue*, Georgia 2004.
Cafaro P., *Thoreau, Leopold, and Carson: Towards an Environmental Virtue Ethics*, "Environmental Ethics" 2011, vol. 23, no. 1, p. 3–17.
Coeckelbergh M., *Environmental Skill. Motivation, Knowledge, and the Possibility of a Non-Romantic Environmental Ethics*, New York 2015.
Gilson E., *Wprowadzenie do nauki św. Augustyna*, Warsaw 1953.
Innocent II *De contemptu mundi sive de miseria conditionis humanae libri tres*, 1828, https://archive.org/details/decontemptumund00achtgoog/page/n2/mode/2up (Access: September 12, 2022).
Jedynak S., *Błąd Naturalistyczny*, "Etyka" 1967, vol. 2, p. 289–297.
MacIntyre A., *Dziedzictwo cnoty. Studium z teorii moralności*, tłum. A. Chmielewski, Warszawa 1996.

Rolston III H., *Environmental Virtue Ethics: Half the Truth but Dangerous as a Whole*, in: *Environmental Virtue Ethics*, R. Sandler, P. Cafaro (ed.), Oxford 2005, p. 61–78.

Sandler R., *Introduction: Environmental Virtue Ethics*, in: *Environmental Virtue Ethics*, R. Sandler, P. Cafaro (ed.), Oxford 2005, p. 1–12.

Treanor B., *The Virtue of Simplicity: Reading Thoreau with Aristotle*, "The Concord Saunterer. New Series" 2007, vol. 15, p. 65–90.

Treanor B., *Emplotting Virtue: A Narrative Approach to Environmental Virtue Ethics*, Albany 2014.

van Wensveen L., *Dirty Virtues: The Emergence of Ecological Virtue Ethics*, Amherst 2000.

Zydorek D., *Cielesność człowieka w świetle średniowiecznych przekazów*, "Medycyna Nowożytna: studia nad historią medycyny" 1996, vol. 2, no. 2, p. 121–130.

8 A proposal of a universalistic, positive, and practical environmental virtue ethics

Although environmental virtue ethics is a relatively young discipline, it has already developed three systematic concepts that offer aretological reflection in the area of humans' ethical obligations to the environment. Each of these concepts is a valuable element in the discussion and has made a valuable contribution to the development of EVE. In this part of the book, I will present a universalistic, positive, and practical concept of EVE, which could serve as a sort of supplementary voice in a multitude of approaches to environmental virtue ethics.

8.1 A universalistic, positive, and practical environmental virtue ethics

My conception of environmental virtues ethics enriches the existing debate by adding features of virtues that have been overlooked in previous analyses. My goal is to draw attention to their importance in the development of environmental moral character. The important feature of virtues that I want to emphasize is their universalistic, positive, and practical character. In this part of the monograph, I will present the meaning of these terms and how they are understood in the context of the aretological discussion of the human–environment relationship. The EVE concept described in this book strongly emphasizes the practical dimension of virtues. Nevertheless, a certain novelty in my approach is the attention paid to reflection on the practical dimension of virtues in language. Hence, in previous sections of this book I devoted considerable attention to Louke van Wensveen's postulate of a return to virtue language in order to propose in this section that this postulate should be corrected. In my opinion, the language of virtues must also describe the practical character of moral proficiencies. On the theoretical side, I rely here on the ideas of Mark Coeckelbergh, whose thought has not been taken into account in previous environmental virtue ethics debates.

The absence of Mark Coeckelbergh's ideas is somewhat understandable. This brilliant philosopher was mainly interested in the philosophy of technology, hence his monograph on environmental skills/competencies (2015) may have escaped the attention of EVE representatives. However, it is worthy of inclusion in the discussion for several reasons. First, his position on virtue is similar to

DOI: 10.4324/9781003433156-12

that of philosophers who consider virtue to be a specific kind of knowledge of how to act.[1] Second, he brings out the practical nature of virtue, in line with the words of MacIntyre (1985): in heroic societies, a person is what it does.

Having presented the framework of my universalist, positive, and practical theory of EVE, I will now present an element without which the practice of virtues would not be possible, namely the sphere of moral education. I believe that every aretological theory is essentially an attempt to put the professed values into practice. Part of the practical sphere is transmission of the cultural heritage to the moral sphere by means of moral education specific to the local tradition. Hence, in addition to ethical theory, the second pillar of EVE is the outline of the idea of moral education, which transforms dry theory into social practice.

8.1.1 The universalistic character of virtue

Universalism in ethics involves the "claim that ethical standards and principles are universal" (O'Neil 1998, 535), therefore lists of virtues, norms, values, and rights are the same for everyone. This claim applies to many ancient traditions as well as to contemporary concepts, such as the concept of human rights. An example of universalism is Cicero's claim that there is one eternal law for all nations and all times (Cicero 1993, 33). I claim that the universalistic character of virtue plays the crucial role in EVE, since to be effective in dealing with ecological crises we need to emphasize what all cultures have in common and what will serve as a factor to unite ourselves in achieving a common goal – overcoming the ecological crisis. In this part of the book, I will present philosophical and empirical arguments for the universal character of values and virtues, as well as examples of the application of the universalism of values and virtues in ecological and educational projects.

8.1.1.1 Philosophical arguments for universalism

I claim that virtues are universal in two senses: (1) human beings share the same understanding of virtues; (2) various cultures consider the same virtues to be the most important. The universal character of virtue is guaranteed by human nature, which requires the same virtues to flourish for all people. At the very general metaphysical level, human beings share a common nature that enables us to recognize the same values and virtues as important.[2] Moreover, the existence of social groups is based on sharing an understanding of the values that are necessary for a given group to exist. John Finnis noticed that how we express particular values may not be the same or universal, and certain values can have different statuses in various cultures, but "the substratum of core values is universal" (Alkire 2002, 172). The universalism of values is possible because the most important values are motivated by basic reasons for activity that are common for all people. These reasons are thought of as areas of realization of well-being and goodness, namely areas where worth and value are gained. They motivate one to act and are universally recognized in various cultures. According to John Finnis (1980, 86–92), the areas of realization of well-being and goodness are life, knowledge and aesthetic experience, work and play, sociability (friendship), practical reasonableness, and religion.

When analyzing Finnis's argumentation on universalism, Alkire gave an example of a situation in which we have an exchange student as a guest who is confused about the concept of a handshake. "She says that she finds it intimidating and is baffled as to what on earth it means" (Alkire 2002, 170). A general explanation will not work here and will leave the student even more perplexed and lost. According to Finnis, we should go to the basic reason that is behind a handshake, because the expression of values may not be universal but the value that is expressed by this gesture is universal. Thus, the explanation that a handshake is a typical way to greet people will not be as effective as explaining the basic reason why we shake hands. According to Alkire (2002), the basic reason for a handshake is the value of friendship. Although the concept of shaking hands as a greeting could be considered exotic, the idea of friendship is common for all cultures. Thus, this foreign guest could agree that friendship is a good reason to act in such an unusual way. According to Finnis's concept, if we keep on explaining more of our actions to the foreign student, we will realize that all our actions relate to a small number of reasons. These reasons exemplify values that are universal for all cultures, and as such they are understandable for people of various cultural backgrounds.

8.1.1.2 *The empirical argument*

In addition to the preceding analysis, I will refer to empirical research that demonstrates that humans share values across cultures. The best known are those proposed by Shalom Schwartz, who has claimed that a society's values are categorized in one of ten groups of values that are shared among cultures. Here, I present the universalism of value, not of virtue, but this needs some explanation. The terms 'values' and 'virtues' are often treated as synonyms or used interchangeably, especially when it comes to practical implementation of moral education (Cf. T. Lovat et al.). For example, in educational programs the term 'values' is sometimes even used in the meaning that is attributed to 'virtues' (Fitzgerald 2014, 137). Even though these two terms have different meanings in the field of philosophy than in other areas, this is not the rule. I present Schwartz's discussion as an interesting example of psychological discussion on the universalism of human nature and shared values. Even though the term 'values' has a different meaning in psychology than in philosophy, it should be emphasized that, in a very general sense, Schwartz's definition of values is similar to the communitarian understanding of virtues.

Schwartz defines values as

> desirable transsituational goals of varying importance that serve as guiding principles in the life of a person or other social entity. Implicit in this definition of values is that (i) they serve the interests of some social entity,

(ii) they can motivate action, giving it direction and emotional intensity, (iii) they function as standards for judging and justifying action, and (iv) they are acquired both through socialization to dominant group values and through individuals' learning experiences.

(Schwartz 1994, 21)

This researcher asked around 60,000 people from different countries and with various cultural and religious backgrounds about their value preferences. Each person was given a questionnaire with a list of 56 potential values and was asked to choose the most important by assigning them numbers (on a 1–7 scale) to indicate how much a given value was appreciated. Subsequently, respondents went through this list of 56 values and rated on a 1–7 scale how each value fares as a guiding principle in one's life (Alkire 2002, 170).

On the basis of these empirical analyses, Schwartz distinguished a set of 11 universal human values: (1) self-direction (independent thought and choice of action); (2) stimulation (e.g., novelty, excitement, and challenges in life); (3) hedonism (related to the need to satisfy sensual pleasures); (4) achievement (the personal need for success); (5) power (e.g., social recognition, wealth, social power, protecting one's public image, authority); (6) security (represents the need for safety and stability); (7) conformity (understood as restraint from activities that upset or harm others); (8) tradition (respect for the traditions and symbols respected in one's culture); (9) spirituality (Schwartz emphasizes that spiritual values are not a guiding principle for everyone)[3]; (10) benevolence (pro-social values that are a cornerstone of good relations in social groups); and (11) universalism (e.g., understanding, tolerance, and preservation of the goodness of people and of nature) (Schwartz 1992, 5–11).

Schwartz is not a fundamentalist in terms of the universalism of values as he does not claim that every person or culture necessarily has these values. Neither does he say that all the aforementioned values are equally important; however, he claims that the values that everyone has as their guiding principles are "understandable in relation to one or another of these categories" (Schwartz 1992). His research provides a very general overview of the shared values that are important across different cultures; it is very informative and provides empirical proof of the aforementioned philosophical claim regarding the universalism of virtues.

8.1.1.3 Universalism in environmental protection and moral education

My view of the universalism of environmental virtues and values could be validated by not only empirical research but also application. The following example doesn't imply that all virtues are universal, but the fact that people from various cultures have acknowledged the same virtues as crucial for the protection of the planet or for education shows that we are able to admit that certain virtues are universal for people all over the world.

The Earth Charter is an example of an internationally discussed document that is based on values respected by representatives of different cultures, races, and religions. Following the publication of the Brundtland report,[4] a discussion arose concerning the need for a new set of norms that would guide humanity to sustainability.[5] These norms came in the form of the Earth Charter, work on which started two years after the Rio Declaration in 1992. Thus, "in 1994, Maurice Strong (Secretary-General of the Rio Summit) and Mikhail Gorbachev (the last leader of the Soviet Union) launched an initiative (with support from the Dutch Government) to develop an Earth Charter" (Earth Charter 2019). Later, the Earth Charter Commission was founded (1997) to support worldwide collaboration on the document. The Charter was discussed with religious and political leaders and people from all regions of the world. Due to the effort of those involved, it was possible to reach a global consensus on the universal virtues that are important and recognized by all nations and are conducive to sustainable development. "The Earth Charter is now increasingly recognized as a global consensus statement on the meaning of sustainability" (Earth Charter 2019). Its creation took a long time, but the work on the charter showed that representatives of various cultures could propose a common catalogue of values that are crucial for environmental protection.

Another example of the practical application of universal virtues is the virtue education program implemented in Australia. This country is very diverse and its inhabitants have various national, cultural, and religious backgrounds; therefore, the idea of common virtues that constitute a basis for education sounds quite challenging in Australia, but it is still possible. The Australian federal government has introduced an obligatory value education program in schools. This program is based on values that are recognized as universal: care and compassion, doing your best, fair go, freedom, honesty and trustworthiness, integrity, respect, responsibility, understanding, tolerance, and inclusion (Snook 2007, 81). The program consists of a few modules and includes the teaching of values and virtues; educational experts actually emphasize that the program often "unwittingly or implicitly refers to virtues rather than values" (Fitzgerald 2014, 137). This educational program, which is based on teaching shared values and virtues, might become a platform for the unification of multicultural and often antagonistic societies. Teaching universal values can support the integration of societies and help future generations to focus on what is common among humans instead of on what differentiates one nation from another. The program is not only an example of the universalism of values and virtues: it also has a beneficial influence on students and their communities, and the positive results of the program include better "self-management, communicative competence, self-reflectivity, resilience, character and integrity" (Lovat et al. 2011, 128). Moreover, the report confirms

> improved classroom and playground behavior, engagement in schoolwork, students negotiating their own learning, improved conflict management, consideration of others, and increases in acts of honesty. At the same time, there have been positive changes in teacher–student relationships, and the

more supportive learning environment has led to growth in mutual respect between teachers and students, including more positive and constructive approaches to behavior management.

(Lovat et al. 2011)

It can be seen that a well-designed moral education program is beneficial for students and their communities and can play a significant role in integrating fragmented and diverse societies. The universalistic aspect of virtue could play a tremendous role in this aspect.

Universalism in EVE is crucial for emphasizing that virtues and values can unify antagonized societies. Thus, it can serve as a tool for uniting societies to overcome the ecological crisis. Universalism applied to virtues could be conducive to moral education founded on the idea that we share the same virtues that are fundamental to our personal flourishing, our *eudaimonia*. Since this will limit their application, we cannot be effective in dealing with ecological problems if we associate virtues with only one religion, creed, or nation. In fact, the association between virtue and religion (their preachy character) has been blamed for the exclusion of this term from ethical discourse (van Wensveen 2000, 6). So, we need to emphasize the environmental virtues that are universal. Some interesting proposals are offered in EVE literature by Ronald Sandler (2007), Brian Treanor (2014), or Philip Cafaro's (2004) interpretation of David Thoreau (2006).

8.1.2 *The positive character of virtue*

The positive character of virtue is a great motivational factor at a personal level. In this part of the chapter, I want to emphasize that an aspirational vision of human excellence works as a positive factor that motivates environmental actions much more effectively than negative input. A lot of environmental activism is based on an apocalyptic scenario of a future natural catastrophe, but the psychological research I present in the following shows that negative messages do not motivate one to act as much as positive ones. Thus, a positive input as a motivational factor is very important at a personal level in order to take individual action and to motivate others to do the same.

Notwithstanding, it should be emphasized that negative scenarios might be an important factor in environmental protection as most of them have informative value.[6] For example, researchers' reports, such as those published by the Intergovernmental Panel on Climate Change, the European Environmental Agency, or the US Environmental Protection Agency, use the prognostic function of science to analyze the current situation and forecast the future state of nature. These scientific reports have high value as an honest source of reliable information about the current situation. Thus, one should not think that scientific prognoses should be ignored or rejected due to their pessimism; rather, they should serve as a tool to recognize what should be done. They have high informative value about the state of nature, and they provide a crucial epistemological background for future actions, even if psychological research shows they do not motivate as strongly as positive input.

8.1.2.1 *The heuristic of fear*

Hans Jonas's (1984) concept of a heuristic of fear is still the philosophical background for fear-based narratives in environmental protection. According to Jonas, fear serves as a trigger for moral agency (Morris 2013, 137) and we should be worried about the terrible future effects of technological activity and be afraid that we will lose our planet. Fear of future *malum* (badness) pushes us to act against it and attempt to save ourselves from future possible destruction. Being responsible for nature necessarily involves fear.

Why fear? There are at least two reasons that fear is so important in Jonas's philosophy. One is that it has agential power: it makes us act to escape what we are worried about. Jonas was against Ernst Bloch's philosophy of hope, which expresses the utopianism of past philosophical theories, of which Jonas was critical. Bloch's utopia foretells goodness ahead while ignoring danger that is quite real. This approach can be summed up by the words 'not yet.' "S is not yet P (the subject is not yet its predicate)" (Jonas 1984, 199). Something is yet to become its predicate – it is not yet what is meant to be; so, the future is a relevant ontological category, and the present matters less than a future that has not yet come. There is some hidden eschatology in awaiting the future while ignoring the present, but we do not know what will happen in the future. Thus, expecting only goodness is the worst-case scenario since we are focusing on something that is neither real nor certain, and it will only allure us with a utopian vision that might not come to pass. Fear does not promise future goodness: it informs us about reality and alerts us so we are able to act. Thus, according to Jonas, to overcome the lethargy created by hope for a better future world, we should apply fear as a factor that encourages us to act against future disaster or loss.

Another argument for fear posits that it is direct and has agential power, while goodness is subtler. Goodness is a subtle witness that might go unnoticed and is a matter of taste, as what is good and appreciated by one might be ignored and not recognized as good by another. Goodness does not provoke action. Jonas claims that

> this is the way we are made; perception of *malum* is infinitely easier for us than perception of *bonum*; it is more direct, more compelling, less given to differences of opinion or taste and, most of all, obtruding itself without our looking for it
>
> (Jonas 1984, 27)

Thus, *malum* (fear) serves as a great tool for environmental protection. The new ethic of Hans Jonas functionalized fear as a cause for environmental action. How do we mobilize action for nature? Jonas provides two objectives of this new ethic.

The first objective of Jonas's new ethic for the future is to visualize the long-term effects of technological enterprise (Jonas 1984). We do not yet experience future badness, but at this stage we should feel the fear of future destruction or losing nature.

> When we are not able to predict the long-term consequences of our actions, he (Jonas) argues that we should proceed with prudence, even to the extent of being guided by fear, in order to ensure that we do not create extensive future harms.
>
> (Morris 2011, 44)

The second objective of Jonas's new ethic is that experience of fear must be visualized: the imagined *malum* should change into the experienced *malum*. This experience will not come on its own, so the task of ethics is to make a moral agent experience genuine fear.

> Ethics should make this distant and imagined evil a real experience for one. The fear (…) cannot be, as in Hobbes, of the 'pathological' sort (to use Kant's term), which compulsively overcomes us in the face of its object, but rather a spiritual sort of fear which is in a sense the work of our own deliberate attitude
>
> (Jonas 1984, 28)

Fear shouldn't paralyze us or push us into pessimism; it should rather act as a tool that keeps us alert and ready to act for the goodness of future generations.

8.1.2.2 *Fear as a factor that deprives one of agency*

There are many environmental programs that functionalize fear. Many prognoses foretell a future that is threatened or even catastrophic. We live in a time of ecological fear that uses the dialectics of "ordinary catastrophe" (Davis 1998). These catastrophes are often analyzed in biblical terms as a sort of apocalypse that has yet to come (Bińczyk 2018, 138–1390. Christian Schwägerl writes that good motives for protecting the environment are accompanied by a "negative attitude, (…) (namely) a certain tendency toward apocalyptic views" (2014, 73). This attitude is especially prevalent in Europe and North America. There are often claims that we are on the verge of an irreversible catastrophe – that we have no more time and soon it will be too late to escape ecological disaster. And then nothing happens. This 'doomsday clock' was set during the cold war:

> this clock had a legitimate role during the Cold War, but since 1989 it seems as if the scientists involved have been striving to invent new reasons to keep their project going instead of donating the clock to a museum.
>
> (Schwägerl, 2014)

This apocalyptical style of narrative is designed to make people act in ways that change the situation.

Do these terrifying pictures and prognoses change anything? No. We can only observe the fetishization of future catastrophe. We enjoy ourselves experiencing visions of future catastrophe as we enjoy a horror movie in the cinema. Schwägerl writes that "something in us loves an Apocalypse – maybe it's the narcissistic feeling of being a member of a chosen generation" (Schwägerl, 2014, 74). According to him, we use fear as a factor for achieving our selfish goals; for example, politicians use it to achieve their aims, and businesses use fear of future catastrophe to make profits. Fear has been functionalized for the realization of certain aims, but it is not useful at all when it comes to solving ecological crises as it does not motivate people to take action; instead, it evokes negative emotions that do not prompt any action. Human nature (Goleman 2013, Chapter 14; Goleman 2019) is such that we do not react to negative input that informs of an impending threat. We are not able to react to threats that we cannot see directly or that are too large to deal with individually, such as climate change or biodiversity loss. Moreover, the vision of a future climate catastrophe doesn't push us to take action because it is too large to deal with. This future disaster is simply too huge, so we ignore negative information about it.

Moreover, we are used to the picture of future ecological catastrophe that has accompanied us for a long time. Thus, fear does not work. Fear has no agential power. Fear will not lead to change; however, it can even cause the opposite change, namely people might think they should enjoy their last days if there is going to be an inescapable future catastrophe. If there is nothing to come after us, then let us use all of the resources and make waiting for the catastrophe an interesting experience. Let us use everything that can be used. Schwägerl claims that this kind of short-term thinking is especially visible in the United States, where "belief in an imminent apocalypse is actively fostered by Evangelical Churches" (Schwägerl, 2014). Thus, fear of a future apocalypse does not support actions that would prevent it. Fear works the other way around.

Should we then reject all ecology of fear and Hans Jonas's concept? Well, we can distinguish two levels: a political one and personal one. At the political level, fear could be a very useful tool, and political programs should focus on the most pessimistic scenarios in order to encourage radical action. Humanity will not survive without strong and decisive political programs to mitigate the results of the destruction of nature and prevent future ecological catastrophes. However, at a personal level, fear has no motivational value. This claim is supported by the empirical research presented in the next section of this book. Thus, the positive aspect of virtue is crucial at the personal level, while fear could be useful at the political level.

8.1.2.3 Why is positive input crucial?

Are we destined to die in a catastrophe that will surely come? No, we can use an input that works – a positive input. In the following I present empirical research results that show that a positive input has greater agential power in terms of environmental actions. A positive input is one that pushes us to act. Thus, the positive and aspirational picture presented by the ideal of virtue can serve as a good motivator and can play an important role in environmental protection.

Research carried out in psychology presents compelling evidence that positive inputs can achieve better results than negative inputs.

> For example, merely renaming a choice to avoid negative associations can make an impact on people's decisions. Saplakoglu and colleagues found that airline passengers were far more willing to pay a surcharge to combat climate change if the fee was called a 'carbon offset' instead of a 'carbon tax'.
>
> (Saplakoglu, 2019)

The term 'tax' has negative connotations; thus, people are less willing to pay for their emissions. While the term 'offset' is neutral, Weber's research has shown that people are more willing to pay for their emissions when this term is used.

Weber claims that positive input plays a crucial role in environmental decision-making because, unlike negative input, it makes people act. The effectiveness of positive input is well established in empirical research into how emotions influence pro-environmental actions. In Weber's research, two possible emotions were taken into consideration: pride and guilt. The results of this research point to a more beneficial role of anticipated pride compared to guilt in shaping pro-environmental motivation and furthermore highlight the need for careful assessment of communication and messaging strategies that employ emotional appeals, as effects may vary substantially depending on the emotion targeted. Policy makers, advocacy organizations and others would benefit from a more nuanced understanding of the impact of induced anticipated emotions on pro-environmental decision making and motivation, to leverage positive effects and avoid potential negative ones. (Schneider et al. 2011)

These results suggest that positive input plays a much greater role in stimulating pro-environmental actions than negative input.

The ecological handprint represents a positive input that is the opposite of the ecological footprint, which is a negative input. The latter informs us how much we have destroyed or used, while the former offers an aspirational view of how much goodness has been done and how positive its impact on a local community is.

> Elke Weber, a cognitive scientist at Columbia University's Earth Institute, says the footprint might be a major reason so few people go from awareness of global warming to ongoing action. When folks harp on the harm we do to the planet we feel bad and want to do something to feel better – and then we tune out. But if we have a positive goal in mind that we can take small, manageable steps toward, we feel good – and so are more likely to keep going.
>
> (Goleman 2013)

The ecological handprint stimulates pro-environmental actions by employing small, simple actions and involving other people in them. The basic idea is to engage new people in pro-environmental activity more effectively than can be achieved with a help of a negative input (see Kühnen et al. 2019). The handprint is widely used in a lot of sustainability education programs (see, Gorana and Kanaujia 2017).

Thus, the presented research proves that positive input is much more effective in terms of motivating pro-environmental actions. Does this mean we should ignore negative input (including all the information about planetary boundaries)? As presented earlier, research shows that negative inputs do not play a motivational role, and Schwägerl emphasizes that the vision of a future apocalypse has no agential power. However, reports about the state of ecosystems have informative value and as such they might be very important. We should constantly monitor the state of the environment to determine what actions should be taken and with what intensity.

In terms of positive input, virtues could play a huge role. Namely, environmental virtues offer the kind of aspirational ideal that people aim for. EVE helps to develop what is good in us; this good will not only help us flourish but is also beneficial for our environment. EVE emphasizes an aspirational vision of personal excellence, and empirical research tends to demonstrate that positive input works better than negative input. With the aid of proper moral education, we will be able to present environmental virtues in the form of daily practices that support environmental protection, and we may achieve more than would be achieved by presenting all possible future disaster scenarios. Virtues have positive power: the power to aspire to be a better person – to be the best version of oneself. This power needs to be harnessed to change approaches toward nature, and EVE could be conducive in that.

8.1.3   *The practical character of virtue*

The third quality of virtue I present is its practical character, which is already included in most virtue ethics (see, Annas 2011, Hursthouse 1999, Foot 2001, Slote 1995). However, the novelty I add to the discussion is the idea of revising the language of virtue so that it reflects the practical character of virtue. Moreover, I focus on moral education that is supported by the concept of narrative. It should be emphasized that virtue is never solely a theoretical concept: it is always more about exercising our positive character traits than it is just about theory. Nonetheless, it should be remembered that "contemporary academic philosophy has lost sight of this important practical goal" (Treanor 2014, 23). According to Treanor, the practical goal of philosophy is simply to live a better life. Academic ethics is far too complicated for a nonacademic to follow. "Many philosophers (…) focus on minute and esoteric details of interest only to other specialists or write in a manner so plodding and obscure as to make their work inaccessible to a general educated audience" (Treanor 2014). Ethics, especially virtue ethics, should be more about the practice, not only about the theory. Thus, sophisticated academic discourses should be accompanied with a discussion that is understandable for nonacademics if they are to be conducive to dealing with ethical challenges in daily life.

Virtue is probably the most applicable concept of ethics. Annas (2011, 16) points out that Aristotle found a similarity between virtue and skill. According to this ancient philosopher, both are practical and can be learned only through practice. As the Stagirite wrote, "we learn by doing; for example, we become builders by building and lyre players by playing the lyre. So too, we become just by doing just actions, temperate by doing temperate actions, and courageous through courageous actions" (Aristotle 2000, a 32–b2). A crucial role in learning virtue is played by the role models that are used in virtue ethics to show what virtues need to be developed and how they should be expressed. Role models are an important element in EVE as well. According to Ronald Sandler, learning by observing a role model is one method of learning what environmental virtue is (Sandler 2005, 5). Exemplars of environmental excellence serve as masters that should be observed and followed – an exemplar is not just to be copied.

> The learner needs to understand what in the role model to follow, what the point is of doing something this way rather than that, what is crucial to the teacher's way of doing things a particular way and what is not. A learner who fails to do this will simply copy the teacher's mannerisms and style along with the teacher's exact way of doing things
>
> (Annas 2011, 17)

Learning a virtue is a specific type of learning that includes understanding what is to be done in a certain situation, therefore part of it is answering the following question: what would the exemplar have done in this situation? It is not simply copying the behavior of the other but acting in a way that includes the lessons from the exemplar's behavior in the learner's own deeds.

The practical aspect of EVE is reflected by virtue language, in the theoretical discussion of which a huge contribution was made by Mark Coeckelbergh and his idea of environmental skills, which is described in detail in Section 8.2. Virtue theory should be accompanied by adequate moral education. Since virtue is the subject of learning and habituation, it is necessary to make it a conscious and deliberate process. Brian Treanor emphasizes the role of narrative, which is genuinely a very useful tool in the personal development of moral agency. Practical application of virtue to moral education is presented in Section 8.3.

8.1.4 Summary

I claim that discussion of environmental virtue ethics should include the three qualities of virtue: its universal, positive, and practical character. I claim that a universalistic perception of virtue is necessary to deal with the ecological crisis. It should be emphasized that we share the same values across cultures. All over the world, we need the same things to achieve well-being, and we require the same values and virtues to achieve human excellence. I support my arguments with empirical research that presents a list of ten shared cross-cultural values (by Shalom Schwartz) and practical experience in elaborating universal catalogues of values for the Earth Charter and moral education. The experiences that have accompanied the preparation of virtue education programs in highly diverse societies such as Australia show that virtues are universal because even diverse nations such as this can work out a catalogue of virtues that is appreciated by all their citizens. This provides proof that the same values and virtues are shared even though cultural, racial, or religious backgrounds might be different. The Earth Charter is another example of an internationally elaborated catalogue of shared values and virtues. I emphasize the universalism of virtues since this is the quality that is crucial for the unification of antagonized societies. Thus, it can serve as a very effective tool for overcoming the ecological crisis. We should focus more on the universal nature of virtues and emphasize what unites societies as this will help enable people from different nations to deal with the global environmental crisis.

The second important aspect of virtue is its positive character, which is visible in aspirational views of virtue as leading to human excellence. In this chapter, I present the argument that positive input is far more effective than negative input in the context of motivating constructive environmental actions. This positive aspect of virtue is a very strong agency factor at the personal level, whereas fear could serve as a tool at the political level. Thus, the prevailing negative discourse about a future ecological apocalypse does not encourage constructive individual actions; instead, it paralyzes moral agents. What is needed at the personal level are positive inputs and virtues that offer an aspirational vision of human excellence. This vision plays a positive role very well.

The third quality of virtue is its practical character. Being virtuous is acting in a virtuous way. Moral education is conducive to supporting the practical character of virtue, but to escape its preachy character the concept of non-preachy narrative is very useful since it helps to tell us about virtue in a subtle way. Moreover, it helps a moral agent to aspire to be a good person (presented in Section 8.3). The third aspect of the practical character of virtue is virtue language. In 2000, van Wensveen claimed that the term "virtue" is not present in environmental ethics discussions.[7] Thus, it should be included there as part of a discussion on the moral obligations of human beings to nature. Her claim was one of the triggers for the genesis of EVE. However, almost two decades later one can notice that virtue terms do not cover all moral challenges and are not practical enough. Thus, I claim that it is necessary to include practical terms such as environmental skill (as was suggested by Mark Coeckelbergh) in the EVE discussion (described in Section 8.2). This concept reframes our thinking about virtues and presents it from a practical, applicable perspective.

8.2 The revised postulate to return to the language of virtues

A key aspect of ecological aretology is the sphere of action and the question of the acts undertaken by a moral agent who is motivated by a specific environmental virtue. The return to the language of virtues postulated by van Wensveen has shaped the discussion of EVE. *Dirty Virtues* (2000) is the first monograph devoted entirely to the issue of environmental virtues, including, to a large extent, questions of virtue language. In the monograph, the author argues for the restoration of virtue language to the environmental debate and the use of the term 'environmental virtue' in place of the terms 'attitude,' 'habit,' or 'practice.' Van Wensveen's monograph revealed the richness of aretological reflection and the multitude of discussions that take place in the field of moral fitness analysis. Almost two and a half decades after the publication of *Dirty Virtues*, it can be seen that this Dutch researcher's postulate requires some revision. EVE needs both language of virtues and terminology that captures the practical dimension of ecological virtues. In the theoretical aspect, van Wensveen's postulate works perfectly well, but in the realm of *praxis* there is a gap between what we know and how this knowledge should be applied. This gap, in my opinion, is filled by Mark Coeckelbergh's concept of environmental skills/competencies (2015).

Coeckelbergh starts with the claim that many people know what is happening to the environment and how destructively it is affected by human actions, yet the actions of these people take are either inadequate or very ineffective. A huge gap is emerging between awareness of environmental degradation and the ability to take ethical action to respond appropriately to the challenges of the environmental crisis. The demand to include environmental literacy in environmental discourse fills this gap. Moreover, in my opinion, removing philosophical terms such as 'attitude,' 'habit,' and 'practice' from environmental discourse impoverishes the ethical debate. The discussion of moral attitudes is an important part of the contemporary aretological debate. When we abandon the use of this terminology, we abandon some important ethical work. To enrich the virtue discussion, Mark Coeckelbergh proposes introducing the terms 'environmental skills' or 'environmental competence' into EVE discourse.

According to Coeckelbergh, Romantic and Enlightenment thinking about nature permeates environmental discourse; it is what prevents us taking action and constitutes a dualism between the natural world and technology (Coeckelbergh, 2015, 4). He claims that restoring the balance between *techne* and *logos* could be achieved through environmental skills/competencies (Coeckelbergh, 2015, 5). These are what make it possible to move from Heidegger's (1962) being-in-the-world to acting-in-the-world. According to Coeckelbergh, such an endeavor makes it possible to stop thinking about the environmental crisis as a process going on somewhere 'out there,' far away from us. With environmental skills/competencies, facing the environmental crisis becomes a duty for the moral agent. This shift in thinking allows us to see the crisis of values and to take genuine responsibility for the problem.

8.2.1   *The influence on environmental discourse of the Enlightenment and Romantic view of nature*

Enlightenment reason and Romanticism feelings weaken the possibility of taking action that could genuinely serve to protect the environment. According to Coeckelbergh, modern environmental philosophy has its origins in Enlightenment philosophy, as is visible in the highly regarded role of science and the approval of the manipulation of nature (Coeckelbergh, 2015, 43–44). Even some ethicians trust science so much that they build their philosophical argumentation on its claims. For example, quite a few analyses begin with (or refer to) the presentation of scientific facts, and this is particularly evident in the case of discussions devoted to issues of climate change or animal welfare.[8] At the same time, our interference in nature with the help of technology creates problems. Thus, we first create a problem and then, with the help of technology, try to solve it, unfortunately without success (Coeckelbergh, 2015, 123). According to Coeckelbergh, the inspiration for modern environmental protection is the same as for the technological intervention in the environment that environmentalists so criticize. This inspiration is Francis Bacon's utopian vision of humans gaining dominion over the world through science. Environmentally minded people, according to Coeckelbergh, are driven by the idea of using knowledge to create a new ecological world in which nature is something external that we must study, protect, and change, hence the constant attempts to manage and manipulate natural resources – to literally construct and create nature – which seemingly stem from the fear that it cannot be left alone (Coeckelbergh, 2015, 45–46). Coeckelbergh gives numerous examples of discussions about restoring wildlife in North America that are always supported by extensive scientific argumentation but also involve calls for changes on a gigantic scale. This philosopher also mentions controversial ideas for restoring species from the very distant past, such as the Pleistocene. To the list compiled by Coeckelbergh one can add a whole range of projects concerning climate engineering, transhumanism, or nanotechnology. All of these are supported by the conviction that humans know best what needs to be done to make the environment function properly, and that human interference in the natural course of things is somehow a necessity. This attitude raises many moral questions.

Enlightenment rationality is not bad in itself; what is undesirable is so-called instrumental rationality (Coeckelbergh, 2015, 52), which is what made reason a controlling rather than a liberating tool. Coeckelbergh uses Max Horkheimer and Theodor Adorno's (2002) concept of humans' control over other people and extends this concept to include humans' control over nature (2002, 53). Moreover, humans since Descartes have been treated as an entity separate from the natural world – alienated from the environment. Coeckelbergh believes that the idea of being outside the natural world is very harmful; he himself is a proponent of Heidegger's idea of 'being-in-the-world.' In this attitude, he sees an antidote to the problem of treating nature as something alien. Such a change in the way we think about nature allows us to become part of the world, from which we were previously separated. At the same time, through this change of perception of nature, we open ourselves up to 'acting-in-the-world,' the idea of which is the cornerstone of environmental virtue ethics as it allows us to move from the concept of virtues as moral ideals to the concept of virtues as moral proficiencies actualized in the daily life of the moral agent.

Heidegger's way of thinking about the moral agent as part of the world prompts a reformulation of the terminology used in environmental protection. Tim Ingold (2000, 20), influenced by Heidegger's thought, talks about the necessity of using the term 'environment' instead of 'nature.' He points out that 'his' environment is an entity analyzed in relation to himself, but at the same time it is subject to certain changes along with himself, as both he and the environment are constantly evolving. Also, the environment is never finished, total, and complete. Above all, however, unlike nature, it is his own authentic environment – it is something in which a moral agent exists. The term 'nature' is scientific and objective; it applies to surroundings to which the moral agent does not belong.

According to Coeckelbergh, cold, rational thinking about nature has failed the test of conservation. Enlightenment rationality as a hidden assumption of conservation discussions did more harm than good. Partial antidotes to extreme rationalism were philosophies, such as German idealism, that emphasized the importance of nonrational factors. Romantic thinking became part of German idealism. A special place in Romanticism was given to immanent goodness and the concept of nature understood as beauty untainted by civilization. Could this type of thinking, then, be a counterweight to Enlightenment rationalism and its legacy? According to Coeckelbergh, definitely not. He is equally critical of viewing the world in a Romantic way, which he considers harmful and not conducive to environmental protection in practice.

Since Romanticism, nature has become the opposite of technology and has been viewed as a source of beauty, value, and goodness. The fascination with the natural world and the attribution of immanent goodness to it in modern philosophy has been present since the time of Rousseau (cf. 1992), who is even referred to as the father of Romanticism. Coeckelbergh (2015, 65) notes that the first distinguishing feature of the Romantic trend was the quest for authenticity – the desire to focus on oneself and turn away from society. This can be achieved in contact with nature. This ideal also underpinned Thoreau's work and thus has influenced environmental philosophy.

Philosophers of technology stress that Romanticism was characterized by a skeptical attitude toward technology (cf. Dusek 2011, 33–43). "Romantics believed that the industrial revolution and new technologies were destroying both nature and the human spirit" (Dusek 2011, 197). Technology was repeatedly demonized in literary works, such as those of William Blake, who wrote about "dark satanic mills," and Thoreau, who strongly condemned civilization, while nature was seen as a miraculous panacea for technology's destruction of the world and spirit. At times nature was even exaggeratedly glorified; for example, Wordsworth writes about one spring breath becoming the source of humans' wisdom (Dusek 2011).

This negative attitude seems to serve environmental protection, but Coeckelbergh sees this approach as a threat to real environmental action. Romanticism is a continuation of thinking based on faith in scientific facts – as was typical of the Enlightenment – that "flows in the veins (of Romanticism)" (Coeckelbergh 2015, 75). The biggest problem with the Romantic way of thinking is that it establishes and sanctions dualism between humans and nature, between technology and humans. This dualism limits our ability to deal with the environmental crisis. This dualistic thinking can be seen in our perception of health; healthy is what is natural, as opposed to what is artificial, which is created by humans and technology. According to Coeckelbergh, it is important to remember that technology does not alienate us from the world; it does not in itself divide the reality around us into what is natural (i.e., good) and what is artificial (i.e., worse than natural). Technology does not alienate us from the surrounding reality but makes us experience it in a different way. Being-in-the-world, postulated by Heidegger, requires redefining dualistic thinking and understanding that technology is one of the means by which we experience being-in-the-world. It is not a source of alienation but rather the cause of a specific way of experiencing the world – a different way of being in it.

Coeckelbergh (2015, 83) rejects Romanticism as much as the Enlightenment. First and foremost, both value the autonomy of the individual. Rationalists believe that science and technology help the individual to be more autonomous, while Romantics argue that science and technology deprive the individual of autonomy. Moreover, Rationalists see the social dimension as an artificial creation that Rationalists want to project and change through engineering, while Romantics reject it because it is a denial of the natural. Coeckelbergh believes that both philosophies make the wrong assumptions. He considers both to be detrimental to environmental protection. In their place, he proposes rethinking our position in the world in light of Heidegger's view. According to Coeckelbergh, this will bridge the gap between theory and action, between knowing that we are experiencing an environmental crisis and knowing what to do to protect the environment.

8.2.2 *Overcoming dualism*

Coeckelbergh devotes much attention to understanding the role of technology in human life. Following Heidegger, he rejects the Enlightenment's idea of mastering the environment through technology but sees the latter as a tool in the world, of which humans are a part. The key to empowering environmental ethics is to move away from human–nature or human–technology dualisms. This type of thinking alienates us from what we are part of. It is important to remember that there is no separate world that is 'nature' somewhere outside of us. Nor is there a technology that is independent of the way we perceive the environment and how we function in it (2015, 87). Coeckelbergh refers to the issue of valuing natural entities, technology, or any elements of our environment. Any valuing we do is only done from a human perspective; even if we have good intentions, we could never take the perspective of a mountain (cf. Leopold 1949, 114–118) or a bat (cf. Nagel 1974). It is humans who judge and value the world around them, and the human perspective is the only one we know (cf. Parker 1996, 33). Coeckelbergh does not see this phenomenon as a problem, unlike many environmental ethicists who find anthropocentrism in this attitude (cf. Minteer 2008, 58–62). He emphasizes that valuing from a human perspective does not mean assigning new values but only discovering and observing the values that have already been created through our active relations with the environment. This way of looking at things allows us to see humans as a part of the environment rather than as alienated entities who – somewhere from a distant perspective – look at it and make value judgements about the world or its individual elements. This contributes to overcoming the dualism that arises when adopting the attitude of an external observer of the world.

Overcoming dualisms can be helped by moving away from using the term 'nature' and replacing it with the term 'environment.' Such a move has great significance for the way humans' place in the world is perceived. First and foremost, the term 'environment' is more inclusive: it takes into account a number of spheres of the moral agent's functioning and makes it possible to include both natural and artificial (i.e., human-made). At the same time, the concept of 'environment' refers not to something that is outside but to what is our immediate environment, connected to us by numerous relations. Coeckelbergh makes use of Merleau-Ponty's concept of the "lived body, which actively relates to things, people, and places" (2015, 88; cf. Merleau-Ponty 1945). With this, the environment should be understood as an intentional entity connected to the perceiving body, as something that is filled with the meaning, power, and sensations of the body. This also means that the environment is dependent on the perceiving subject; it is a space for the perceiving subject to experience the world – to be in the world – rather than an alien external dimension that we observe from a certain distance.

This kind of shift in thinking has an impact on our perception of our relationships and obligations to the environment. By understanding that we are connected to the environment, we know that, in order to be a better person, we cannot leave the environment when it is threatened.

8.2.3 *Environmental skills – from being-in-the-world to acting-in-the-world*

The rejection of thinking that alienates human beings from the environment and deprives environmental discourse of its power to have a real impact on the environment is the first step toward a change of thinking in environmental ethics. After the shift from nature to environment, the second important thing is closing the gap between knowing about the environmental crisis and acting to overcome it. In this respect, Coeckelbergh sees environmental virtue ethics as a way from being-in-the-world to acting-in-the-world. He starts his reflection by defining EVE (2015, 5). First, he rejects the Stoic view of virtue; in particular, he is critical of its overemphasis on the virtues of moderation and limitation, which are rooted in the Stoic view that material goods are unimportant and the only thing important to man is the development of virtues. Second, as noted earlier, Coeckelbergh rejects the Romantic heritage of environmental thinking. Third, he abandons the Aristotelian interpretation in which great emphasis is put on theoretical knowledge.

According to Coeckelbergh, the practical dimension of Aristotle's thought is crucial. Merely knowing what is good is not enough to do good. To do good, environmental skills/competencies are necessary as they help one to move from theoretical knowledge to practical application of it. With environmental skills/competencies, we can be genuinely environmentally virtuous as we not only know what is happening to the environment, but we also have the knowledge of how to act in order to realize environmental virtues. Coeckelbergh claims that environmental skills/competencies allow us to transform abstract knowledge into authentic life experiences and commitment to the environment (2015, 99). In doing so, these skills bridge the gap between factual knowledge and practical knowledge. This is a contribution to giving environmental virtue ethics a new quality.

Applying environmental skills/competencies helps us not only to protect the environment but also to become better persons. According to Coeckelbergh, this practical aspect was present in earlier concepts of EVE. In Thoreau, it can be found in the form of the virtue of skillful action. Thoreau (2006), in describing his experiences at Lake Walden, gives a number of practical tips on how to act in various spheres of human life, including quite practical recommendations on resource management, diet (linked to praise of vegetarianism), growing vegetables, dealing with certain animals, and respect for the natural environment.

Coeckelbergh's concept may be particularly useful for environmental virtue ethics; in his view, virtues are not abstract principles or rules. Virtues require the practical involvement of a moral agent who knows what to do. Thus, the virtue of skillful action is crucial. Coeckelbergh emphasizes that his ethics of skillfulness understands being virtuous not only as a matter of acquiring intellectual knowledge, but also as developing practical skills and engaging with the environment (natural and social), that is, moral knowledge of what to do. Environmental virtues, therefore, are not merely a character trait or external action but concern the way we act in the world, for they influence our relationship with the environment – our being-in-the-world (2015, 5).

The virtue of skillful action is only partly inspired by Aristotle's virtue ethics. According to Coeckelbergh, of key importance here is the emotional factor. This philosopher believes that environmental emotions[9] push us to protect the environment. Above all, forming and nurturing environmental virtue requires engaging and developing practical and environmental skills (2015, 123). Virtue cannot only be about knowledge detached from reality: it must include knowledge of how to act.[10] According to Coeckelbergh, theoretical knowledge detached from practice is not only not a solution but is itself a problem. It is a bad guide on the road to environmental protection as it only creates the appearance of involvement in activities that benefit our natural environment, deceiving the moral agent and unjustifiably soothing his conscience.

The ethics of skill is inseparable from activities in the world, such as walking (the other activity valued by Coeckelbergh is working in nature). It was walking that Romanticism's forefathers gave special importance in literature. However, in connection with the rejection of certain Romantic categories, the concept of walking in nature should also be rethought. Romantics had the idea of escaping from noisy and polluted cities to the bosom of nature, uncontaminated by human activity. However, as was pointed out earlier, the Romantic concept of nature alienates the moral agent from himself as it presents culture as something alien to a moral agent. Walking integrates both the cultural and the biological parts of humans; it is exercise for the body and spirit (2015, 138): the body – through the physical effort put into moving the muscles while walking; the spirit – by exercising a kind of harmony, striving for unity with the natural world. Walking may also have scientific purposes, such as gathering information about the natural environment and its elements, about certain plants or animals.

This meaning of walking is part of a Romantic narrative, that is, one that assumes a dualism between an action (a symbol of a certain culture) and the culture that the action reflects. Overcoming this dualism, according to Coeckelbergh, is possible by appealing to the category of skill. Coeckelbergh cites the views of Tim Ingold and Lee Vergunst (2008), who describe walking as an activity that constitutes a way of thinking and feeling. Walking is therefore part of both nature and culture. It is part of a practice; or, more accurately, a practice and experience that is as much natural as it is cultural (2008, 139). In other words, it is an activity in which the entire human is involved. Walking is not just a way of moving but an activity that gives the moral agent the opportunity to feel his own body and the natural environment in which the body resides. We have the possibility of experiencing being-in-the-world, establishing a relationship with the world, and experiencing it through the body and mind.

Sometimes while walking we assist ourselves with technologies that help us find our way, such as GPS navigation. However, by directing his attention to tracking the direction of movement on an electronic medium, this distracts the moral agent from connecting with the environment. Coeckelbergh defends the use of maps and navigation. He is consistent in opposing antagonizing the natural world and technology. In his view, using technology to navigate should be viewed not negatively but as a certain kind of experience. This experience does not make contact with nature cease to matter but only changes the way we experience the world around us. The use of technology is an example of the use of practical reason. It is a fact that the world of technology makes experiencing the world different. Technology is connected to our bodies, but it does not cut us off completely from experiencing the world.

8.2.4 Summary

Van Wensveen's proposal to revisit the language of virtues in discussions of environmental ethics played an important role in the emergence of environmental virtue ethics. First of all, her publication was a strong impetus for the emergence of several of the concepts of environmental virtue ethics discussed in this section. Time has shown that, for various reasons, the very phrase 'environmental virtue' does not reflect the richness of philosophical reflection. In my view, talking about moral dispositions by means of classical philosophical terminology narrows the debate because it overlooks many phenomena of environmental virtue ethics. Therefore, I believe that discussions of environmental virtue ethics should be broadened to include philosophical terms that capture the practical dimension of virtues; the category of environmental skills/competencies proves extremely useful here. A key argument in favor of expanding the language of virtues is that the phenomenon of virtues cannot be encapsulated solely in the terms 'virtue' and 'vice.'

8.3 Moral education as one of the pillars of the universalist, positive, and practical concept of environmental virtue ethics

The essence of virtue is its practical[11] dimension. Since the time of Aristotle, virtue has been linked to the sphere of action, not just to ethical theory itself. In my opinion, the theory of virtue ethics acquires value only in conjunction with education. The practical dimension of environmental virtue ethics requires tools to help it emerge from the framework of theory alone. Hence, it can be said that EVE is like a building set on two pillars: one is theory, and the other is moral education, which is the most effective tool for allowing theory to become social practice. In this way, hermetic ethical discourse can go beyond its own framework and serve social change better than the most sophisticated theory.

8.3.1 The social dimension of virtue

According to Treanor (2014, 156), virtue ethics is the heart of ethical formation because it teaches the moral agent how to move from the question "Who am I?" to the question "Who should I be?" Hursthouse claims that virtue ethics is an ethical concept that addresses the transformation from who we were to who we should be (2019). She emphasizes that according to Aristotle our moral decisions depend on the moral formation we underwent in childhood. At the same time, moral education is much more complex than simply following a catalog of commands and prohibitions. Treanor (2014, 156) writes that it would have been infinitely easier for him as a parent to instill in his daughters the maxims of Enlightenment projects of normative ethics, such as "actions are right insofar as they promote happiness" (Mill, 1979, 7), or "act only in accordance with that maxim through which you can at the same time will that it become a universal law" (Kant 1997, 31). Virtue ethics, as I wrote about previously, often criticizes ethics built on the model of legal systems because they are not simple to apply. Sometimes the complexity of a situation requires more elaborate analysis, and a specific catalog of principles cannot be so easily applied. The complexity of reality is not compatible with codes of precepts and prohibitions offered by normative ethics. Moreover, new ethical challenges continually arise; for example, the development of science and technology and the growing human impact on the natural environment mean that nature should be the subject of our moral choices.

Any set of precepts and prohibitions will be no more than a record on paper if it is not linked to the moral agent's inclinations to carry them out. Even the most sophisticated norms will prove useless if a moral agent does not have the moral proficiencies necessary to follow them. Thus, virtues constitute morally good acts. According to Treanor, the complexity of moral situations also makes us as adults fail to evaluate our actions in terms of following norms (2014, 166). When we analyze our actions and refer to norms, we most often ask ourselves what a good (virtuous) person would do in this situation. How would a virtuous person act in light of this principle? However, this does not mean that norms are useless and there is no need to refer to them. Norms are crucial as a moral compass, but mere knowledge of even the best catalogs of commands and prohibitions does not guarantee good behavior. Knowledge of norms does not make us good people. Ethics begins with being a good person, that is, internal work, which is a guarantee of the application of moral norms. Moreover, possessing virtues allows one to apply certain moral norms adequately to the situation.

People with poorly developed virtues should apply norms in order to live a good life. Here, the observation of more-experienced people or their advice can be crucial, as long as it is virtuous. Thus, an important element is the so-called social dimension of virtue, that is, appealing to moral models provided by people we perceive as more virtuous than ourselves. The role of a moral exemplar is one of the essential elements of virtue ethics. Even if we are not virtuous ourselves, we can still recognize virtuous people and can model ourselves after them to make the right moral decisions. Thus, the lack of certain moral skills does not spoil our chances of doing good.

Central to moral education is the social dimension of virtue. Treanor stresses that it is the duty of the ethicist to have a broad social impact and influence political decisions to create conditions for the development and nurturing of virtues. He proposes dissecting virtues on three levels: individual, social, and environmental. In this chapter, I will focus on the social and environmental dimensions of virtue. Environmental virtues involve many spheres of a person's life and are related to their functioning in society. Moreover, the cultural and historical context determines how we view virtues.

First, virtues do not develop in a social vacuum. Assigning value to certain virtues is always linked to a specific social and historical context. All rules always apply in a specific society and time. There are no universal principles, and morality is always a morality of society (cf. MacIntyre 1985). This means that moral theories relate to the cultural context and time in which they are created. This does not contradict the universalistic character of virtues.

Their essence remains the same, only the way of reading virtue and its context changes: the core of virtues is universal (cf. Alkire 2002, 172). However, the ways of realization are various and are aligned with the norms of each social group.

The way we talk about virtue is always laden with tradition and is also constantly revised with new ways of reading it that are dependent on changes in the cultural context. Moreover, Max Weber's writings (1999) show that the very model of the hero and the virtues attributed to him can change. Weber argues that the model of the Homeric warrior today is the clerk, but his qualities are very different from his ancient counterpart. A good clerk must first and foremost be obedient. The priority is not courage or justice, but precisely servility to the state apparatus. Weber's analysis is critical of the way society is organized, but when read in the light of MacIntyre's words it can help to see how social change leads to the development of different social roles and practices and thus to the creation of catalogs of virtues specific to certain social groups.

In addition, some moral standards that are valid in one era may become controversial in other times; for example, revenge was an important category in some pre-Christian sagas. Retaliation is seen not as an inability to forgive but as a display of bravery and courage. Revenge itself is an interesting area for cultural analysis; in the early Middle Ages, for example, German law provided for punishment for murder only if it was the secret murder of an unidentified person. "When a known person kills another known person, not criminal law but revenge by a kinsman is regarded as the appropriate response" (cf. MacIntyre 1985, 166). In modern times, revenge for murder seems almost as punishable as murder itself. All acts of independent judgment, especially those resulting in someone's death, are punishable. While acting in retaliation may serve as a mitigating circumstance, it does not remove the guilt of the one who carries out the revenge. This change in attitudes toward revenge-driven killing shows a change in the way this act is viewed and is due to changes in morality dictated by the cultural and historical context.

Every practice involves certain patterns of excellence to which the moral agent must conform. Moreover, practices have a history that is shaped by a particular cultural and historical context. The history of the development of each social practice is itself more than just the history of the technical means used in it (MacIntyre 1985, 193). MacIntyre emphasizes that practice (in which virtues are situated) is not a set of technical skills: while every action requires a certain set of technical skills, what distinguishes practice is its reference to the intrinsic and extrinsic goods that characterize the moral agent. Thus, framing virtue in terms of practices does not contradict what has been written here about environmental skills.

8.3.2 The importance of education in environmental virtue ethics

Becoming virtuous requires the exercise of virtue (Annas 2011, 12) and this is what moral education offers us. Virtues are acquired through education, habituation, and exercise (Annas 2011).

> What we need to learn to do, we learn by doing; for example, we become builders by building, and lyre players by playing the lyre. So too we become just by doing just actions, temperate by doing temperate actions, and courageous by courageous actions
>
> (Aristotle 2000, 1103a32)

Thus, according to Annas, ethical education is a necessary part of virtue theory; it is not 'merely practical' and not only sidelined to theory. We cannot understand what virtue is without understanding how to acquire it (cf. Annas 2011, 21).

The theoretical discussion provides answers to questions regarding moral theory, but only the practical application of this theory is the litmus test for the validity of the beliefs proclaimed. The aspect of the transmission of virtues and their practice is particularly emphasized by Brian Treanor as his whole concept is subordinated to the idea of passing on virtue to the widest possible audience in a way that inspires its practice. Nevertheless, the inseparability of the idea of virtue from its practice can be seen in each of the theories discussed. In Thoreau's case, it takes the form of the moral agent's individual journey of discovering himself and striving continually to become a better version of himself. The harsh conditions of the New England woods become Thoreau's private Kurukszetra, where his virtues attempted to overcome his vices.

Moral education is the key to introducing environmental virtues into social practice and is a necessary complement to EVE theory. Indeed, environmental virtue ethics requires a two-step approach: first, the development of a specific theory that determines how to understand the basic categories of virtue ethics; second, it is necessary to complement this theory with an aspect of moral education that will enable the moral agent to attain the exemplary moral character defined by ethical theory. Contemporary virtue ethicists recognize the role of moral education and "encourage the improvement of moral character through the acquisition of ethical virtues" (Szutta 2017, 100). It is emphasized, however, that moral education does not relate to a behavioral understanding of virtue and is therefore not merely the practice of behavior appropriate to a particular situation. Virtue ethicists accept the premise that virtue is not merely the mechanical performance of learned actions. True virtue involves several cognitive-affective processes (Szutta 2017, 146), that is, processes that also relate to practical wisdom (phronesis) and require of the moral agent a certain kind of awareness and emotional maturity. The intellectual factor is crucial: according to MacIntyre (1985), usage of intelligence is even the basic feature that distinguishes a natural disposition to a certain kind of virtue. Thus, moral education would face the challenge of shaping the moral agent on various levels. Its aim would be to help the individual to develop practical wisdom and to prepare them holistically to fully develop their own moral faculties. But is moral education, which aims to help the individual to develop his or her moral faculties, possible at all?

The first professional teachers of virtue were the sophists, who in a sense broadened the scope of *arete* by making the privilege of possessing it available to everyone, not only the nobly born, and offering all the wealthy the opportunity to acquire the skills necessary to function well in the *polis* (Szudra-Barszcz 2010, 109–110). What they taught still did not imply moral prowess but showed great faith in education's capacity to teach specific skills to anyone interested. The belief in the possibility of learning virtue became a subject of criticism by Socrates, who in his dialogue with Protagoras exposes the superficiality of sophistic teachings. The sophists' teachings aroused much controversy because of their adoption of a relativistic conception of truth and their belief that they could convince disputants to adopt their views, regardless of whether these views were true or only served the sophist's own interests.

According to Socrates, there is no way to teach someone virtue because the shaping of virtue is a matter of self-formation, and part of shaping it is understanding and knowing what it is. Even a virtuous person cannot force someone to form virtue in themselves.[12] Aristotle formulates a different view on the possibility of teaching virtue. He recognizes that it is insufficient to merely be aware of what virtue is and that it is crucial to know how to practice it. Aristotle takes the view that virtues can be taught, especially ethical virtues. The very process of teaching virtues is part of the formation of human beings. Virtues can be formed through positive action, by improving moral skills, or through negative action by combating one's own weaknesses (Szudra-Barszcz 2010, 113).

According to Szutta, virtue in the behavioral approach seems to be easier to shape, as the process of an individual's moral formation resembles physical training. It is a kind of training of a "moral muscle" (2017, 150). The element of invoking exemplary moral character, cited earlier, refers precisely to this type of virtue training. A moral exemplar helps the moral agent decide what a moral hero would do in a given situation. Virtue in cognitive-affective terms implies the learning of appropriate cognitive functions (phronesis) as well as affective functions (mainly appropriate motivation). In this second form of education, literature plays a huge role as an appropriate narrative that helps to develop desirable moral attitudes (cf. Treanor 2014; Nussbaum 1990a, 1990b).

It is crucial to start this type of education at the earliest stages of education. Bonnie Kent (1999) draws attention to the words of Aristotle, who believed that adequate habituation should begin in early childhood, a condition for the formation of good moral character in later life. The formation of moral character requires the tools of communication to be adapted to the age of the recipient, therefore the moral education of children differs from that of adolescents or adults. However, in the case of character education through narrative, the task seems to be easier in the case of children because children's literature abounds in a number of positive role models and characters who can serve as moral role models – it is enough to take care to select the right content. One form of narrative is the personal narrative that helps us give meaning to our actions; it helps us understand the events in our lives. "Thus, we use narrative (…) to understand life retrospectively" (Treanor 2014, 181) and understand it in a bigger context. Gare claims that "the narrative of one's life is part of an interrelated set of narratives, defining the individual as a member of a number of interrelated institutions: a family, a city, a guild or profession, different organizations, and a nation" (Gare 1998, 5).

Personal narrative is important not only for oneself: it can serve as a source of inspiration for others. It can be seen as a very vivid example of ethical beliefs, as is the case with Thoreau's *Walden* (Cafaro 1999, 109–110). This book is an example of a sort of self-creation where each and every act is understood in the wider context of self-development in union with nature. *Walden* is a very personal journey in which the main character explores his authentic self and develops a moral character. Thus, Thoreau's book has made a significant impact on environmental ethics. As Cafaro notices, "such a philosopher may lead both by example and by powerful personal testimony. These are often stronger forces for change than arguments from principles" (Cafaro 1999, 109). Thoreau's narrative is also part of a wider discussion. He had knowledge of issues such as the economy, but he explains them in a way that is part of his own very personal history, his own flourishing, and the message he wants to convey. In this way, ethical theory is described as Thoreau's experience, and this makes the theory a vivid, authentic concept.

Personal narrative can be source of inspiration for others, as is the case with Thoreau. However, one never lives in a societal vacuum; we are determined by tradition and the society we live in. Gare sees the individual's narrative as strongly determined by tradition. He sees narrative as a part of successful life and acquiring virtues. He claims that "as a members of communities, individuals are engaged in the quest for a successful life" (Gare 1998, 5), to which virtue is conducive. Virtue is understood as a disposition "which sustains practices and enables people to achieve the goods internal to those practices and, most importantly, sustain people in the relevant kind of quest for the good" (Gare 1998, 6). What is important is some sort of idea of what the ultimate good actually is since this knowledge helps in understanding what kind of character is conducive to achieving this good. This helps us to understand which character traits or virtues should be acquired to achieve this good. The personal narrative is supportive here and is strongly determined by the tradition one is part of.

Tradition influences the way we understand and perceive the world, including our beliefs about the natural environment (White 1973, 57). Thus, the proper narratives can influence the way in which the environment is perceived. The aim of environmental philosophy is to change the way the environment is perceived.

> A new ethics, nevertheless, even if focused on the narratives people are living out, their practices, the virtues they uphold and their ideas of the good, will not be enough in isolation from other discourses. It is also necessary to overcome the parlous state of philosophy, which is no longer able to resolve disputes about ethical issues or any other issue
>
> (Gare 1998, 8)

Brian Treanor's concept of narrative environmental virtue ethics recognizes the role of narrative in shaping character traits and beliefs. He sees literature as a great source of narrative that supports the formation of moral character. Treanor's concept of EVE shows how literature can support moral education and personal flourishing.

Along with the demand for moral education in terms of environmental virtues, the question of how attitudes should be formed remains to be resolved. While detailed decisions should be left to specialists in the field of moral education, I believe that several elements have emerged in philosophy that are valuable for educators. First, the consideration of moral dilemmas can be helpful as it draws attention to certain ethical problems, sensitizes the moral agent to them, and helps them to find solutions. Second, I think narrative can be very helpful as it is a subtle tool that is also fascinating because it can draw on a huge number of literary works, folk messages, legends, or myths.

8.3.3 Summary

The practical nature of virtues requires tools that support their practice. In addition to such changes being made to the language of virtues to accommodate the practical sphere, appropriate moral education is also necessary. Virtues are not innate, but moral education is an indispensable tool for nurturing them. Its purpose is to form moral agents so that they strive for their own moral perfection. Virtue ethicists emphasize that moral education should become part of virtue theory because becoming virtuous itself requires perfecting oneself in virtue. Adequate moral education is the litmus test of virtue theory. I believe that the next stage of work on environmental virtue ethics should be to propose moral education programs in this area. Thus, moral education can become a tool for increasing environmental awareness.

Notes

1 Mark Coeckelbergh (2015) refers to virtue as know-how, literally "operational knowledge." This expresses the practical dimension of virtue, but this practical dimension is often framed in a reductionist way that overly simplifies virtue and presents ethics from the perspective of empirical research (see, for example, Dreyfus, Dreyfus, 1990; Churchland 2000; Clark 2000a, 2000b).
2 In the book, I adopt the essentialist view in this regard; this perspective is often a metaphysical foundation of universalism in ethics and can be traced back to Aristotelian philosophy.
3 This is why some authors who have analyzed Schwartz's research mention only ten universal values.
4 A report entitled *Our Common Future* was published in 1987 by The World Commission on Environment and Development. Its publication was a major steppingstone in the introduction of the concept of sustainable development.
5 Even if most of the discussions on sustainability remain anthropocentric, this is a form of anthropocentrism that could be called prudential or enlightened as it acknowledges the privileged position of human beings in some sense but emphasizes their moral obligations toward nature. It is not an arrogant form of anthropocentrism that justifies exploitation of resources.
6 In the first version of my idea, I rejected negative images as not conducive to action. However, discussion with Zbigniew Wróblewski led me to adopt a less radical position and acknowledge the informative value of negative discourse.
7 However, it should be emphasized that many authors had written about environmental virtues before 2000.
8 More on relations between environmental ethics and environmental science can be found in *Environmental Ethics. Theory in Practice* (2018).
9 Environmental emotions are understood here as feelings that a person has for the environment and/or elements of it. These states are an important stimulus in taking action to protect the environment. The canvas for this position is David Hume's views on the role of emotions in morality.
10 According to Coeckelbergh, environmental skills are 1. Health skills. The ability to take care of one's health while being aware of being part of the environment in which one functions; 2. Food and eating skills. The ability to consume food appropriately, such as according to the concepts of the slow food movement, the ideas of which allow us to better engage with the environment; 3. Animal relations skills. The ability to think of animals as part of our world and care for them. Food ethics is linked to animal ethics because a relational connection to the world requires abolishing the antagonism between the human and animal worlds and rethinking human–animal relationships; 4. Energy skills. The ability to choose the right sources of energy, and the ability to use this resource rationally; 5. Climate change skills. The ability to deal with climate change. Being-in-the-world requires involvement in the problems that occur in it, including the issue of climate change. Skillful action cannot be viewed along the lines of managing the climate and trying to control these changes from a global perspective. Skillful action in this area means a series of local actions that are needed in a given area (Coeckelbergh 2015, 157–171).
11 In some sense, my approach aligns with environmental pragmatism, which argues that theoretical debates should be combined with practical engagement, mostly political and social, and activism. In Section 8.3, I focus only on moral education, and my limited approach to practical activity does not fully reflect the richness of environmental pragmatism's claims. However, some common beliefs regarding the need for practical application of knowledge are shared by my approach and environmental pragmatism (Cf. Katz and Light 1996; Minteer and Manning 1999; Minteer 2012; Wenz 1997).

12 Brian Treanor (2014, 160) draws attention to the role of inspiration for becoming
    virtuous. So, while a virtuous person cannot teach virtue to another person, he can
    play a significant role in inspiring someone's personal transformation.

References

Alkire S., *Global Citizenship and Common Values*, w, *Global Citizenship, A Critical
    Reader*, N. Dower, J. Williams (ed.), Edinburgh 2002, p. 169–182.
Annas J., *Intelligent Virtue*, Oxford 2011.
Aristotle, *Nicomachean Ethics*, trans. R. Crisp. Cambridge 2000.
Bińczyk E., *Epoka człowieka. Retoryka i marazm antropocenu*, Warsaw 2018.
Cafaro P., *Personal Narratives and Environmental Ethics*, "Environmental Ethics" 1999,
    vol. 20, no. 1, p. 109–110.
Cafaro P., *Thoreau's Living Ethics. Walden and the Pursuit of Virtue*, Georgia 2004.
Churchland P., *Rules, Know-How and the Future of Moral Cognition*, "Canadian Journal
    of Philosophy" 2000, vol. 26, no. 1, p. 291–306.
Cicero, *De Republica*, Harmondsworth 1993.
Clark A., *Word and Action: Reconciling Rules and Know-How in Moral Cognition*,
    "Canadian Journal of Philosophy" 2000a, vol. 26, p. 267–289.
Clark A., *Making Moral Space: A Reply to Churchland*, "Canadian Journal of
    Philosophy" 2000b, vol. 26, p. 307–312.
Coeckelbergh M., *Environmental Skill. Motivation, Knowledge, and the Possibility of a
    Non-Romantic Environmental Ethics*, New York 2015.
Davis M., *Ecology of fear. Los Angeles and the imagination of disaster*, New York, 1998.
Dreyfus H. L., S. E. Dreyfus, *What is Morality? A Phenomenological Account of the
    Development of Ethical Expertise*, in: *Universalism vs Communitarianism*,
    D. Rasmussen (ed.), Cambridge 1990, p. 237–264.
Dusek V., *Wprowadzenie do filozofii techniki*, transl. Z. Kasprzyk, Cracow 2011.
Earth Charter, *History of the Earth Charter*, https://earthcharter.org (Access: June 25,
    2019).
Finnis J., *Natural Law and Natural Rights*, Oxford 1980.
Fitzgerald M., *Master thesis on Virtues and optimal moral education in the values educa-
    tion in Australian schools' project*, School of Philosophy and Theology, The University
    of Notre Dame 2014.
Foot P., *Natural Goodness*, Oxford 2001.
Gare A., *MacIntyre, Narratives, and Environmental Ethics*, "Environmental Ethics"
    1998, vol. 20, no. 1, p. 3–21.
Goleman D., *Focus, The Hidden Driver of Excellence*, New York 2013.
Goleman D., *Ecological intelligence*, https://www.danielgoleman.info/handprints/
    (Access: August 20, 2019).
Gorana R. N., P. R. Kanaujia, *A New Paradigm of Education Towards Sustainable
    Development*, in: *Reorienting Educational Efforts for Sustainable Development,
    Schooling for Sustainable Development*, R.N. Gorana, P.R. Kaaujia (ed.), Dordrecht
    2017, p. 21–34.
Heidegger M., *Being and Time*, trans. J. Macquarrie, E. Robinson, Oxford 1962.
Horkheimer M., T. Adorno, *Dialectic of Enlightenment*, transl. E. Jephcott, Stanford
    2002.
Hursthouse R., *On Virtue Ethics*, Oxford 1999.
Hursthouse R., *Virtue Ethics*, in: *Stanford Encyclopedia of Philosophy*, E.N. Zalta (ed.)
    (Winter 2016 edition), https://plato.stanford.edu/archives/win2016/entries/ethics-
    virtue (Access: June 25, 2019).
Ingold T., *The Perception of the Environment: Essays on Livelihood, Dwelling and Skill*,
    London–New York 2000.

Ingold T., J.L. Vergunst, *Ways of Walking: Ethnography and Practice on Foot*, Aldershot 2008.

Jonas H., *The Imperative of Responsibility, In Search of an Ethics for the Technological Age*, Chicago 1984.

Kant I., *Groundwork of the Metaphysics and Morals*, transl. M. Gregor, Cambridge 1997.

Katz E., Light A., *Environmental Pragmatism*, New York 1996.

Kent B., *Moral Growth and the Unity of the Virtues*, in: *Virtue Ethics and Moral Education*, D. Carr, J. Steutel (ed.), New York 1999, p. 113–128.

Kühnen M. et al., *Contributions to the Sustainable Development Goals in Life Cycle Sustainability Assessment, Insights from the Handprint Research Project*, "Nachhaltigkeits Management Forum" 2019, vol. 27, no. 1, p. 65–88.

Leopold A., *A Sand County Almanac*, London 1949.

Lovat T. et al., *Values Pedagogy and Student Achievement*, Dordrecht 2011.

MacIntyre A., *After Virtue*, London 1985.

Merleau-Ponty M., *Phénoménologie de la perception*, Paris 1945.

Mill J. S., *Utilitarianism*, Cambridge 1979.

Minteer B. A., *Anthropocentrism*, in: *Encyclopedia of Environmental Ethics and Philosophy*, J. B. Callicott, R. Frodeman (ed.), Farmington Hills 2008, p. 58–62.

Minteer B. A., *Refounding Environmental Ethics: Pragmatism, Principle, and Practice*, Philadelphia 2012.

Minteer B. A., R.E. Manning, *Pragmatism in Environmental Ethics*, "Environmental Ethics" 1999, vol. 21, no. 2, p. 191–207.

Morris T., *What It means to Be Responsible. Reflections on Our Responsibility for the Future*, "Theoretical & Applied Ethics" 2011, vol. 1, no. 2, p. 42–47.

Morris T., *Hans Jonas's Ethic of Responsibility, From Ontology to Ecology*, Albany, NY 2013.

Nagel T., *What Is It Like to Be a Bat?*, "The Philosophical Review" 1974, vol. 83, no. 4, p. 435–450.

Nussbaum M., *Form and Content, Philosophy and Literature*, in: *Love's Knowledge. Essays on Philosophy and Literature*, Eadem, M. Nussbaum (ed.), Oxford 1990a.

Nussbaum M., *Reading for Life*, in: *Love's Knowledge. Essays on Philosophy and Literature*, M. Nussbaum (ed.), Oxford 1990b.

O'Neil O., *Universalism in Ethics*, in: *Routledge Encyclopedia of Philosophy*, E. Craig (ed.), vol. 9, New York 1998, 535–539.

Parker K., *Pragmatism an Environmental Thought*, in: *Environmental Pragmatism*, A. Light, E. Katz (ed.), New York 1996, p. 21–37.

Rousseau J.J., *The Reveries of the Solitary Walker*, transl. C.E. Butterworth, Indianapolis 1992.

Sandler R., *Introduction: Environmental Virtue Ethics*, in: *Environmental Virtue Ethics*, R. Sandler, P. Cafaro (ed.), Oxford 2005.

Sandler R., *Character and Environment. A Virtue-Oriented Approach to Environmental Ethics*, New York 2007.

Saplakoglu Y., *Better Decision-Making for the Planet*, https://www.princeton.edu/news/2018/01/04/better-decision-making-planet (Access: August 12, 2019).

Schneider C., L. Zaval, E. Weber, E. M. Markowitz, *The Influence of Anticipated Pride and Guilt on Pro-Environmental Decision Making*, "PLoS ONE" 2011, vol. 12, no. 11. DOI: 10.1371/journal.pone.0188781

Schwägerl C., *The Anthropocene. The Human Era and How it Shapes Our Planet*, Santa Fe, London 2014.

Schwartz S., *Universals in the Content and Structure of Values, Theory and Empirical Tests in 20 Countries*, "Advances in Experimental Social Psychology" 1992, vol. 25, p. 5–11.

Schwartz S., *Are There Universal Aspects in the Content and Structure of Values?*, "Journal of Social Issues" 1994, vol. 50, no. 4, p. 19–45.

Slote M., *From Morality to Virtue*, Oxford 1995.

Snook I., *Values Education in Context*, in: *Values Education and Lifelong Learning*, D.N. Aspin, J.D. Chapman (eds.), vol. 10, Dordrecht 2007, p. 80–92.

Szudra-Barszcz A., *Czy cnoty można nauczyć?*, "Ethos" 2010, vol. 23, no. 4, p. 108–118.

Szutta N., *Czy istnieje coś, co zwiemy moralnym charakterem i cnotą?*, Lublin 2017.

Thoreau H. D., *Walden; Or, Life in the Woods*, "The Pennsylvania State University (for the source electronic book file version) 2006.

Treanor B., *Emplotting Virtue: A Narrative Approach to Environmental Virtue Ethics*, Albany 2014.

van Wensveen L., *Dirty Virtues: The Emergence of Ecological Virtue Ethics*, Amherst 2000.

Weber M., *Panowanie urzędników a przywództwo polityczne*, tłum. J. Sidorek, in: *Max Weber*, Z. Krasnodębski (ed.), Warsaw 1999, p. 178–198.

Wenz P. S., *Environmental Pragmatism*, "Environmental Ethics" 1997, vol. 37, no. 3, p. 327–330.

White L., *Continuing the Conversation*, in: *Western Man and Environmental Ethics – Attitudes Toward Nature and Technology*, I. G. Barbour (ed.), Boston 1973, p. 55–64.

Conclusion

Environmental virtue ethics attempts to use ancient wisdom to analyze the moral dimension of the human–environment relationship. It stands out from previous discussions in that it analyzes this relationship from an aretological perspective. In this way, it supplements previous discussions with questions about the nature of the moral agent and his obligations to the natural environment. The EVE framework is dominated by inspirations flowing from virtue ethics in Aristotelian terms, but the basic concepts of this philosopher are interpreted in the context of a challenge unknown in Aristotle's time: the environmental crisis. This crisis led to the development of the concept of environmental virtue, which defines the moral dispositions of an individual in the context of his functioning in the natural environment. The way environmental virtues and vices are understood varies greatly. In fact, environmental virtue ethics is an interesting area of research in that, despite being a fairly young discipline, it has developed some very different approaches. The concepts of environmental virtue ethics can be compared to plants that grow in the same soil but each has grown in different conditions, making them different from each other. The soil of each of these concepts is primarily American transcendentalism (mainly the thought of Henry David Thoreau) and Aldo Leopold's idea of a biotic community of life.

Each of the three concepts of environmental virtue ethics that are analyzed in this book grew in the fertile soil of Taylor's biocentrism. At the same time, each was fueled by different intellectual inspirations, therefore all these concepts are different, despite their common ground. The classical concept is the closest to its inspirations, showing Thoreau's figure as an unsurpassed ideal of environmental virtue ethics. It forms an original blend of the eclectic current that was American Transcendentalism with the thought of Aristotle. The guiding principle of this concept is moderation or even an ascetic model of life, combined with the pursuit of moral excellence developed due to unity with nature. Respect for the Noble Old Lady, as nature is referred to, is Thoreau's main motto. By living in harmony with nature, one gets to know it and oneself better and better, in the style of the ancient *gnoti se auton*, or romantic hero. Getting to know oneself helps to discover the worth of Thoreau's third key virtue: freedom. Freedom to which, following Far Eastern inspirations,

DOI: 10.4324/9781003433156-13

modesty and moderation lead. Liberation from possessions and the desire for ownership is the true liberation of the spirit in the Far Eastern style.

While Thoreau looked to Far Eastern thought for inspiration, Ronald Sandler went in a completely different direction, drawing on the views of contemporary virtue ethicists. As an example of the application of virtue ethics to solve the problems of modern humans, this inspiration made his conception part of the mainstream of the renaissance of virtue ethics. The main theses of Sandler's approach are inspired by the thought of esteemed contemporary ethicists: Rosalind Hursthouse, Philippa Foot, and, to a lesser extent, Gertrude Anscombe. Sandler's concept is primarily naturalistic, in the sense that it emphasizes the biological nature of man. At the same time, it does not limit the essence of humanity to biology alone. Hence, this ethics of environmental virtues could be called holistic since it takes into account the importance of biological and non-biological goals in human life. These goals are part of Sandler's ethics and indicate its teleological character, which leads man to achieve eudaimonia, or a happy life. The third significant feature of Sandler's ethics is its pluralistic character, understood as the pursuit of eudaimonic and non-eudaimonic goals. In this way, the moral agent cares for both himself and other entities. These other entities here are both human and nonhuman beings. Sandler thus spells out a maximally wide range of entities covered by moral consideration.

The third systematic and original take on environmental virtue ethics is Brian Treanor's narrative concept of EVE, which was inspired mainly by the thought of Paul Ricoeur and Martha Nussbaum, but also Alasdair MacIntyre. Its most important feature is the practical dimension of environmental virtue ethics, seen primarily in Treanor's postulate of incorporating ethical norms into political decisions, as well as in concern for the proper transmission of environmental virtues. The latter goal is to be realized mainly through narrative, in all kinds of literary works promoting environmental virtues. Narrative spreads knowledge of virtues in a subtle and friendly way and helps the moral agent to shape himself. In this way, it can play a huge role in moral education, which is crucial for developing virtues of all kinds.

Such is the panorama of environmental virtue ethics. It should be noted that a significant impetus for the development of this discipline was Louke van Wensveen's *Dirty Virtues: the Emergence of Ecological Virtue Ethics* (2000). Although this author presented a non-systematic concept of environmental virtue ethics, she was the first to recognize the presence of virtues in discussions of environmental ethics. In doing so, van Wensveen opened the eyes of researchers to the fact that the discussion of environmental virtues has been going on for a long time but lacks a clear indication that it is an aretological debate. It was the publication of van Wensveen's book that brought the topic of environmental virtue ethics to the attention of modern ethicists and gave rise to the three systematic concepts mentioned earlier.

I supplement the previous discussion of EVE by presenting a universalist, positive, and practical environmental virtue ethics. I base my claim of the universality of virtues on the assumed constancy of human nature, which means that the same virtues and values are esteemed throughout time and in different cultures. In addition, I refer to the functional dimension of virtue in the terms of John Finnis's philosophy, which states that the functioning of social groups, societies, and even international communities is based on the same understanding of values. Finnis's analyses have shown that virtues can differ both formally and in how they are expressed, but "the substratum of core values is universal" (Alkire 2002, 172). A special function in the universalism of virtues is played by environmental virtues, which, due to their common object (the global environmental crisis), need an approach that takes into account the universalist nature of virtues. Emphasizing the universalistic character of environmental virtues is a crucial tool in overcoming the environmental crisis.

The second important feature of virtues is their positive nature, understood as their orientation to developing the good in humans and serving their development. This approach is new in environmental ethics as previous ethics focused on negative stimuli, such as the heuristics of fear. In this monograph, I defend the claim that environmental virtue ethics cannot focus on the rhetoric of fear or other negative stimuli. Such environmental strategies have not worked and, in fact, have frozen the moral agent with fear, thus failing to induce him to take any action to protect the environment. I refer here to philosophical analyses showing the role of positive incentives in encouraging action (e.g., Hume, Coeckelbergh). In addition, psychological research confirms that the rhetoric of fear cannot induce people to act because it induces a defensive reflex, that is, the moral agent closes himself off to negative information. Hence, I advocate reaching for the positive moral models conveyed by virtues, in which I see an opportunity to stimulate action much more effectively than through negative stimuli because they show a certain moral ideal to strive for, thus arousing positive aspirations. Moreover, unlike fear, they stimulate action. Thus, it is necessary in environmental ethics to move away from ineffective negative rhetoric and postulate positive incentives, and virtue ethics can be of great help in this regard.

The third important feature of virtues is their practical nature, which is not a novel trait of ethics as many ethicists emphasize the importance of practicing virtues. Nevertheless, this is such an important feature that I could not leave it out of my EVE proposal. Virtues are qualities that we cannot pronounce we have until we apply them in various situations. In this part of the monograph, I present selected philosophical positions affirming the importance of practicing virtues (e.g., Julia Annas, Brian Treanor). It is the importance of the practical dimension of virtues that inspired me to pay attention to two issues: the language of virtues and moral education.

The practical dimension of virtues requires appropriate tools for their implementation. Thus, in the book I justify the thesis that in environmental virtue ethics the language of virtues should be expanded to include philosophical concepts that take into account the practical nature of virtues. I point out arguments for the need to correct van Wensveen's postulate to return to the language of virtues. The discussion of virtues cannot be limited to the terms "ecological/environmental virtue" or "ecological/environmental vice." In my opinion, Mark Coeckelbergh's proposal to speak of environmental skills/competencies captures the practical dimension of virtue well and should be included in the discussion of the moral dimension of environmental protection. The meaning of the word "virtue" has undergone many modifications (which I present extensively in Part One of this monograph). This richness of meaning indicates the vividness of virtue language, which corresponds to the cultural changes in a given society.

As I wrote previously, environmental virtue ethics requires a marriage with appropriate moral education. The practical dimension of environmental virtue ethics needs thought and action to make it a discipline that is practiced on a wide scale, and moral education is the surest way to achieve this goal. Of particular importance is the education of children and adolescents, although all forms of promoting this knowledge among adults are also valuable. Thus, I see the future direction of the discipline mainly in the development of appropriate proposals for the education of environmental virtues through formal and informal education. I believe that it is possible to raise environmental awareness by taking advantage of the positive and universal nature of virtues. Besides, the very practical dimension of virtues requires that the concept of environmental virtue ethics does not remain only in the realm of theory.

Finally, it is worth asking whether environmental virtue ethics can be relevant to solving the problems faced by modern man. The scientific and technological revolution, the progressive technologization of life, or the catastrophic state of many ecosystems means that the moral agent often has limited opportunities to influence the reality around him. Many of the possibilities for solving global problems lie outside the sphere of individual choices. Nonetheless, moral agents' moral competence is crucial in spheres in which they have the opportunity to act and make a difference. In this sense, environmental virtue ethics can be a tool for effectively assisting change and countering environmental degradation.

References

Alkire S., *Global Citizenship and Common Values*, w, *Global Citizenship, A Critical Reader*, N. Dower, J. Williams (ed.), Edinburgh 2002, p. 169–182.

Index

For Product Safety Concerns and Information please contact our EU representative GPSR@taylorandfrancis.com
Taylor & Francis Verlag GmbH, Kaufingerstraße 24, 80331 München, Germany

www.ingramcontent.com/pod-product-compliance
Lightning Source LLC
Chambersburg PA
CBHW060307220326
41598CB00027B/4265